# Graph-Theoretical Matrices in Chemistry

*Mathematics and chemistry make excellent partners.*

**Dennis H. Rouvray**
Editorial Foreword
*J. Math. Chem.* **1** (1987)

# Graph-Theoretical Matrices in Chemistry

## Dušanka Janežič

*University of Primorska, Faculty of Mathematics,
Natural Sciences and Information Technologies,
Koper, Slovenia*

## Ante Miličević

*The Institute for Medical Research and Occupational Health,
Zagreb, Croatia*

## Sonja Nikolić

*The Rugjer Boškovic Institute, Zagreb, Croatia*

## Nenad Trinajstić

*The Rugjer Boškovic Institute, Zagreb, Croatia*

**CRC Press**
Taylor & Francis Group
Boca Raton London New York

CRC Press is an imprint of the
Taylor & Francis Group, an **informa** business

CRC Press
Taylor & Francis Group
6000 Broken Sound Parkway NW, Suite 300
Boca Raton, FL 33487-2742

First issued in paperback 2017

© 2015 by Taylor & Francis Group, LLC
CRC Press is an imprint of Taylor & Francis Group, an Informa business

No claim to original U.S. Government works

ISBN-13: 978-1-4987-0115-0 (hbk)
ISBN-13: 978-1-138-89428-0 (pbk)

**Visit the Taylor & Francis Web site at**
**http://www.taylorandfrancis.com**

**and the CRC Press Web site at**
**http://www.crcpress.com**

# Contents

# Preface to the Second Edition

The first edition of this monograph, entitled *Graph-Theoretical Matrices in Chemistry*, was published, under the editorship of Professor Ivan Gutman, by the University of Kragujevac, Kragujevac, Serbia, in 2007 (Janežić et al., 2007). That monograph appeared in the series Mathematical Chemistry Monographs as the third volume and is sold out. Therefore, we decided to improve and enlarge the second edition of the monograph in order to include novel graph-theoretical matrices that appeared after the first edition was published. We also received quite a few comments on the first edition of the monograph; the most detailed were by Professor Milan Randić, the foremost mathematical chemist of our times. In the second edition of the monograph, we include most of his comments as well as his suggestions to include a few more graph-theoretical matrices from the early days of the chemical graph theory. The term *chemical graph theory* was introduced in early 1970s in the Theoretical Chemistry Group at the Rugjer Bošković Institute in Zagreb. Nenad Trinajstić first used this term that is now generally accepted for chemical applications of graph theory (Gutman, 2003).

Several new monographs appeared since 2007 reporting in part on graph-theoretical matrices and related molecular descriptors, e.g., *Molecular Descriptors for Chemoinformatics* (Todeschini and Consonni, 2009), *Statistical Modelling of Molecular Descriptors in QSAR/QSPR* (Dehmer et al., 2012) and *Mathematical Chemistry and Chemoinformatics* (Kerber et al., 2014). Several new graph-theoretical matrices have also been proposed, such as the sum-connectivity matrix (Zhou and Trinajstić, 2010) or the distance-weighted adjacent matrix (Randić et al., 2010) and the matrix of dominant distances in a graph (Randić, 2013). The total number of graph-theoretical matrices considered here is 170.

The second edition is organized similarly as the first edition; that is, after the introduction the considered graph-theoretical matrices are presented in five chapters "The Adjacency Matrix and Related Matrices," "Incidence Matrices," "The Distance Matrix and Related Matrices," "Special Matrices," and "Graphical Matrices." Each chapter is followed by a list of references. The monograph ends with concluding remarks and a subject index.

We thank Dr. Sc. Bono Lučić for his help with this manuscript by providing reprints of a number of papers we needed to consult. We also thank the master of engineering in landscape architecture Zdenko Blažeković for his help with figures. Comments by reviewers were most helpful. We thank them for their valuable suggestions.

## REFERENCES

M. Dehmer, K. Varmuza, and D. Bonchev, eds., *Statistical modelling of molecular descriptors in QSAR/QSPR*, Wiley-Blackwell, Weinheim, 2012.

I. Gutman, Preface to a special issue entitled *Graph-based molecular structure-descriptors—theory and applications*, *Ind. J. Chem.* 42A (2003) 1197–1198.

D. Janežić, A. Miličević, S. Nikolić, and N. Trinajstić, *Graph-Theoretical Matrices in Chemistry*, Mathematical Chemistry Monographs, MCM-Vol. 3, I. Gutman, Ed., University of Kragujevac, Kragujevac, 2007.

A. Kerber, R. Laue, M. Meringer, Ch. Rücker, and E. Schymanski, *Mathematical chemistry and chemoinformatics*, De Gruyter, Berlin/Boston, 2014.

M. Randić, $D_{MAX}$—Matrix of dominant distances in a graph, *MATCH Commun. Math. Comput. Chem.* 70 (2013) 221–238.

M. Randić, T. Pisanski, M. Novič, and D. Plavšić, Novel graph distance matrix, *J. Comput. Chem.* 31 (2010) 1832–1841.

R. Todeschini and V. Consonni, *Molecular descriptors for chemoinformatics*, Vols. I and II, Wiley-VCH, Weinheim, 2009.

B. Zhou and N. Trinajstić, On the sum-connectivity matrix and sum-connectivity energy of a (molecular) graph, *Acta Chim. Slov.* 57 (2010) 518–523.

# Preface to the First Edition

Mathematical chemistry has a long history extending back to the times of Russian polymath Mikhail Vasilyevich Lomonosov (Oranienbaum, from 1948; Lomonosov, 1711–Sankt Peterburg, 1765), when he attempted in the mid-18th century to mathematize chemistry (Trinajstić and Gutman, 2002). A part of mathematical chemistry that we call chemical graph theory (Trinajstić, 1983, 1992) also has a distinguished past that extends to the second half of the 19th century when Arthur Cayley (Richmond, Surrey, 1821–Cambridge, 1895) was enumerating alkane isomers (Cayley, 1875) and James Joseph Sylvester (London, 1814–London, 1897) introduced the terms *algebraic chemistry* and *graph* (Sylvester, 1877/1878, 1878). Alexander Crum Brown (Edinburgh, 1838–Edinburgh, 1922), who was trained in both chemistry and mathematics, was probably the first chemist who did research in mathematical chemistry (Crum Brown, 1864, 1866/1867).

The term *algebraic chemistry* has in due course been replaced by the more general term *mathematical chemistry*, but a better term than *graph* has never been found. The seminal role of Cayley and Sylvester in the early development of mathematical chemistry in general and chemical graph theory in particular has been expertly reviewed by Dennis H. Rouvray (1989). It is important to point out why mathematical chemistry is relevant to chemistry. We could not do better than Jerome Karle, Nobel Prize Laureate 1985, who wrote: "Mathematical chemistry provides the framework and broad foundation on which chemical science proceeds" (Karle, 1986).

Mathematical chemistry and chemical graph theory were developing sluggishly with only a few leaps, such as Pólya's work on combinatorial enumeration (Polya, 1937), until the 1970s. Then there suddenly appeared several research groups, located worldwide, that started to speedily develop chemical graph theory. One of the directions in which this vigorous revival was moving was the introduction of a number of novel graph-theoretical matrices.

Matrices are the backbone of chemical graph theory. Classical graph-theoretical matrices are the (vertex-) *adjacency matrix*, the (vertex-edge) *incidence matrix*, and the (vertex-) *distance matrix* (Harary, 1971; Behzad and Chartrand, 1971; Johnson and Johnson, 1972; Wilson, 1972; Bondy and Murty, 1976; Rouvray, 1976; Chartrand, 1977; Cvetković et al., 1988, 1995; Buckley and Harary, 1990). Historically, incidence matrices appear to have been the first to be used (Poincaré, 1900). However, the most important graph-theoretical matrix is the vertex-adjacency matrix, as is well illustrated by Cvetković, Doob, and Sachs in their monograph *Spectra of Graphs— Theory and Applications* (Cvetković et al., 1995), the first edition of which appeared in 1982. An important source for the distance matrix is the monograph *Distance in Graphs* by Buckley and Harary (1990).

In the last 25 years perhaps more than 100 novel graph-theoretical matrices have been introduced. Among the literature sources reporting some of these matrices and their uses are the monographs *Topological Indices and Related Descriptors in QSAR and QSPR*, edited by Devillers and Balaban (1999), *Handbook of Molecular Descriptors* by Todeschini and Consonni (2000), and *Molecular Topology* by Diudea et al. (2001), and the review articles "Molecular Graph Matrices and Derived Structural Descriptors" by Ivanciuc et al. (1997) and "Eigenvalues as Molecular Descriptors" by Randić et al. (2001).

We present 130 graph-theoretical matrices in the encyclopedic manner, classified into five groups: adjacency matrices and related matrices, incidence matrices, distance matrices and related matrices, special matrices, and graphical matrices. The motivation for preparing this monograph comes from the fact that among the matrices presented, several are novel, several are known only to a few, and the properties and potential usefulness of many graph-theoretical matrices in chemistry are yet to be investigated.

Most of the graph-theoretical matrices that we present here have been used as sources of molecular descriptors, usually referred to as topological indices—the term *topological index* was introduced 35 years ago by Hosoya (1971)—which have found considerable application in structure-property-activity modeling (Trinajstić, 1983, 1992; Gutman and Polansky, 1986; Devillers and Balaban, 1999; Karelson, 2000; Diudea, 2001; Diudea et al., 2001), usually abbreviated QSPR (quantitative structure-property relationship) (Sabljić and Trinajstić, 1981) and QSAR (quantitative structure-activity relationship) (Tichy, 1976). Graph-theoretical and related matrices, however, have also been used for many other purposes in chemistry (e.g., Randić, 1974; Hendrickson and Toczko, 1983; Lukovits, 2000, 2002, 2004; Lukovits and Gutman, 2002; Klein et al., 2002; Babić et al., 2004; Miličević and Trinajstić, 2006; Diudea et al., 2006) and in other sciences (e.g., Avondo-Bodino, 1962; Johnson and Johnson, 1972; Chartrand, 1977; Hage and Harary, 1986; Roberts, 1989).

Hopefully, this monograph will stimulate some readers to undertake research in this fruitful and rewarding area of chemical graph theory and introduce new kinds of graph-theoretical matrices that may find use in chemistry.

Finally, we wish to point out that this book is an outcome of the long-standing Croatian-Slovenian joint research collaboration in computational and mathematical chemistry.

The authors thank G.W.A. Milne, former editor-in-chief of the *Journal of Chemical Information and Computer Sciences*, for his editorial assistance with this book.

## REFERENCES

G. Avondo-Bodino, *Economic applications of the theory of graphs*, Gordon & Breach, New York, 1962.

D. Babić, D.J. Klein, J. von Knop, and N. Trinajstić, Combinatorial enumeration in chemistry, in *Chemical modelling: Applications and theory*, ed. A. Hinchliffe, Vol. 3, Royal Society of Chemistry, Cambridge, 2004, pp. 126–170.

M. Behzad and G. Chartrand, *Introduction to the theory of graphs*, Allyn & Bacon, Boston, 1971.

J.A. Bondy and U.S.R. Murty, *Graph theory with applications*, North Holland/Elsevier, Amsterdam, 1976.

F. Buckley and F. Harary, *Distance in graphs*, Addison-Wesley, Reading, MA, 1990.

A. Cayley, Über die analytischen Figuren, welche in der Mathematik Bäume gennant werden und ihre Anwendung auf die Theorie chemischer Verbindungen, *Ber. Dtsch. Chem. Ges.* 8 (1875) 1056–1059.

G. Chartrand, *Graphs as mathematical models*, Prindle, Weber and Schmidt, Boston, 1977.

A. Crum Brown, On the theory of isomeric compounds, *Trans. R. Soc. (Edinburgh)* 23 (1864) 707–719.

A. Crum Brown, On an application of mathematics to chemistry, *Proc. R. Soc. (Edinburgh)* VI(73) (1866–1867) 89–90.

D. Cvetković, M. Doob, I. Gutman, and A. Torgašev, *Recent results in the theory of graph spectra*, North Holland, Amsterdam, 1988.

D. Cvetković, M. Doob, and H. Sachs, *Spectra of graphs—Theory and applications*, 3rd ed., Johann Ambrosius Barth Verlag, Heidelberg, 1995.

J. Devillers and A.T. Balaban, eds., *Topological indices and related descriptors in QSAR and QSPR*, Gordon & Breach, Amsterdam, 1999.

M.V. Diudea, ed., *QSPR/QSAR studies by molecular descriptors*, Nova, Huntington, NY, 2001.

M.V. Diudea, M.S. Florescu, and P.V. Khadikar, *Molecular topology and its applications*, EfiCon Press, Bucharest, 2006.

M.V. Diudea, I. Gutman, and J. Lorentz, *Molecular topology*, Nova, Huntington, NY, 2001.

I. Gutman and O.E. Polansky, *Mathematical concepts in organic chemistry*, Springer, Berlin, 1986.

P. Hage and F. Harary, *Structural models in anthropology*, Cambridge University Press, Cambridge, 1986.

F. Harary, *Graph theory*, 2nd printing, Addison-Wesley, Reading, MA, 1971.

J.B. Hendrickson and A.G. Toczko, Unique numbering and cataloging of molecular structures, *J. Chem. Inf. Comput. Sci.* 23 (1983) 171–177.

H. Hosoya, Topological index. A newly proposed quantity characterizing the topological nature of structural isomers of saturated hydrocarbons, *Bull. Chem. Soc. Jpn.* 44 (1971) 2332–2339.

O. Ivanciuc, T. Ivanciuc, and M.V. Diudea, Molecular graph matrices and derived structural descriptors, *SAR QSAR Environ. Res.* 7 (1997) 63–87.

D.E. Johnson and J.R. Johnson, *Graph theory with engineering applications*, Ronald, New York, 1972.

M. Karelson, *Molecular descriptors in QSAR/QSPR*, Wiley-Interscience, New York, 2000.

J. Karle, Letter of welcome (dated October 10, 1986) to D.H. Rouvray on the occasion of the first issue of the *Journal of Mathematical Chemistry*. Physical chemist Jerome Karl (born as Jerome Karfunkle; New York, 1918–Annandale, Virginia, 2013) and mathematician Herbert Aaron Hauptman (New York, 1917–Buffalo, NY, 2011) were awarded the Nobel Prize for Chemistry in 1985 for their work in mathematical chemistry, more specifically for developing the mathematical approach by which they solved the phase problem in x-ray chrystallography.

D.J. Klein, D. Babić, and N. Trinajstić, Enumeration in chemistry, in *Chemical modelling: Applications and theory*, ed. A. Hinchliffe, Vol. 2, Royal Society of Chemistry, Cambridge, 2002, pp. 56–95.

I. Lukovits, A compact form of the adjacency matrix, *J. Chem. Inf. Comput. Sci.* 40 (2000) 1147–1150.

I. Lukovits, The generation of formulas for isomers, in *Topology in chemistry: Discrete mathematics of molecules*, ed. D.H. Rouvray and R.B. King, Horwood, Chichester, 2002, pp. 327–337.

I. Lukovits, Constructive enumeration of chiral isomers of alkanes, *Croat. Chem. Acta* 77 (2004) 295–300.

I. Lukovits and I. Gutman, On Morgan trees, *Croat. Chem. Acta* 75 (2002) 563–576.

A. Miličević and N. Trinajstić, Combinatorial enumeration in chemistry, in *Chemical modelling: Applications and theory*, ed. A. Hinchliffe, Vol. 4, Royal Society of Chemistry, Cambridge, 2006, pp. 408–472.

H. Poincaré, Second complément à l'Analysis Situs, *Proc. London Math. Soc.* 32 (1900) 277–308.

G. Pólya, Kombinatorische Anzahlbestimmungen für Gruppen, Graphen und chemische Verbindungen, *Acta Math.* 68 (1937) 145–254. Translation of this paper appeared in the book form: G. Pólya and R.C. Read, *Combinatorial enumeration of groups, graphs and chemical compounds*, Springer, New York, 1987.

M. Randić, On the recognition of identical graphs representing molecular topology, *J. Chem. Phys.* 60 (1974) 3920–3928.

M. Randić, M. Vračko, and M. Novič, Eigenvalues as molecular descriptors, in *QSPR/QSAR studies by molecular descriptors*, ed. M.V. Diudea, Nova, Huntington, NY, 2001, pp. 147–211.

F.S. Roberts, ed., *Applications of combinatorics and graph theory to the biological and social sciences*, Springer-Verlag, New York, 1989.

D.H. Rouvray, The topological matrix in quantum chemistry, in *Chemical applications of graph theory*, ed. A.T. Balaban, Academic, London, 1976, pp. 175–221.

D.H. Rouvray, The pioneering contributions of Cayley and Sylvester to the mathematical description of chemical structure, *J. Mol. Struct. (Theochem)* 185 (1989) 1–14.

A. Sabljić and N. Trinajstić, QSAR: The role of topological indices, *Acta Pharm. Jugosl.* 31 (1981) 189–214; to our knowledge, the abbreviation *QSPR* was used for the first time in print in this report.

J.J. Sylvester, Chemistry and algebra, *Nature* 17 (1877/1878) 284.

J.J. Sylvester, On an application of the new atomic theory to the graphical representation of the invariants and covariants of binary quantities, with three appendices, *Am. J. Math.* 1 (1878) 64–125.

M. Tichy, ed., *Quantitative structure-activity relationships*, Akademiai Kiado, Budapest, 1976. This book is a collection of papers based on the reports presented at the Conference on Chemical Structure-Biological Activity Relationships: Quantitative Approach, Prague, June 1973. To our knowledge, the abbreviation *QSAR* was used for the first time in print in this book.

R. Todeschini and V. Consonni, *Handbook of molecular descriptors*, Wiley-VCH, Weinheim, 2000.

N. Trinajstić, *Chemical graph theory*, Vols. I and II, CRC, Boca Raton, FL, 1983.

N. Trinajstić, *Chemical graph theory*, 2nd ed., CRC, Boca Raton, FL, 1992.

N. Trinajstić and I. Gutman, Mathematical chemistry, *Croat. Chem. Acta* 75 (2002) 329–356.

R.J. Wilson, *Introduction to graph theory*, Oliver and Boyd, Oxford, 1972.

# 1 Introduction

The aim of this monograph is to present a number of the graph-theoretical matrices and related matrices that are frequently encountered in chemical graph theory. Matrices are convenient devices for the algebraic representation of graphs—they allow numerical handling of graphs (e.g., Randić, 1974; Hendrickson and Toczko, 1983; Lukovits, 2000, 2002, 2004; Lukovits and Gutman, 2002; Klein et al., 2002; Babić et al., 2004; Miličević and Trinajstić, 2006; Diudea et al., 2006). A *graph* is a mathematical object, usually denoted by $G$, which consists of two nonempty sets; one set, usually denoted by $V$, is a set of elements called *vertices*, and the other, usually denoted by $E$, is a set of unordered pairs of distinct elements of $V$ called *edges* (Wilson, 1972). Thus, $G = (V, E)$. Note in the parlance of Harary (1971), vertices are called *points* and edges *lines*. The degree of a given vertex in $G$ is equal to the number of adjacent vertices, denoted by $d$.

We are here concerned with a special class of graphs called *chemical graphs*, that is, graphs representing chemical structures. If chemical structures under consideration are molecules, we call this type of chemical graph a *molecular graph*. They are generated by replacing atoms and bonds with vertices and edges, respectively (Trinajstić, 1983, 1992; Gutman and Polansky, 1986). Hydrogen atoms are ordinarily neglected. A picture of a simple molecular graph $G_1$ representing the hydrogen-depleted carbon skeleton of 1-ethyl-2-methylcyclobutane is given in Figure 1.1.

A *simple graph* is defined as a graph that contains no multiple edges or loops. Two or more edges that join a pair of vertices are called *multiple* edges. A graph containing multiple edges is called the *multiple graph* or *multigraph* (Harary, 1971). A *loop*

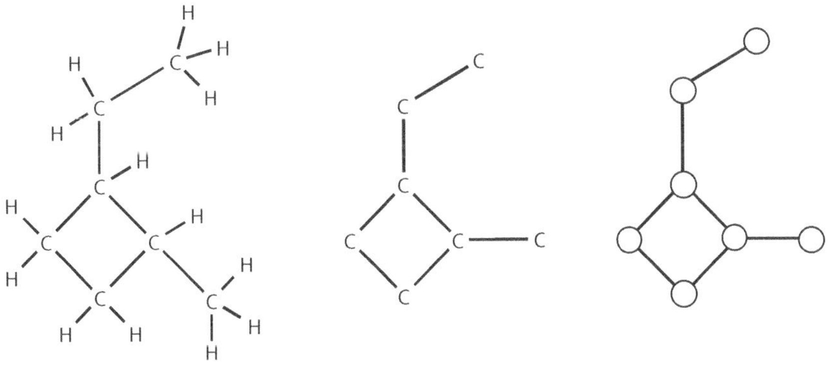

**FIGURE 1.1**  1-Ethyl-2-methylcyclobutane $C_7H_{14}$, its hydrogen-depleted carbon skeleton $C_7$, and the corresponding molecular graph $G_1$.

is an edge joining a vertex to itself. Graphs containing multiple edges and loops are called *general graphs* (Wilson, 1972).

Labeling vertices and edges of a graph is important, because the structure of any graph-theoretical matrix depends on the labeling (Trinajstić, 1983, 1992). In other words, two graphs may be identical, but because they are differently labeled, the corresponding matrices will appear to be different in their manifested arrangements.

## REFERENCES

D. Babić, D.J. Klein, J. von Knop, and N. Trinajstić, Combinatorial enumeration in chemistry, in *Chemical modelling: Applications and theory*, ed. A. Hinchliffe, Vol. 3, Royal Society of Chemistry, Cambridge, 2004, pp. 126–170.

M.V. Diudea, M.S. Florescu, and P.V. Khadikar, *Molecular topology and its applications*, EfiCon Press, Bucharest, 2006.

I. Gutman and O.E. Polansky, *Mathematical concepts in organic chemistry*, Springer, Berlin, 1986.

F. Harary, *Graph theory*, 2nd printing, Addison-Wesley, Reading, MA, 1971.

J.B. Hendrickson and A.G. Toczko, Unique numbering and cataloging of molecular structures, *J. Chem. Inf. Comput. Sci.* 23 (1983) 171–177.

D.J. Klein, D. Babić, and N. Trinajstić, Enumeration in chemistry, in *Chemical modelling: Applications and theory*, ed. A. Hinchliffe, Vol. 2, Royal Society of Chemistry, Cambridge, 2002, pp. 56–95.

I. Lukovits, A compact form of the adjacency matrix, *J. Chem. Inf. Comput. Sci.* 40 (2000) 1147–1150.

I. Lukovits, The generation of formulas for isomers, in *Topology in chemistry: Discrete mathematics of molecules*, ed. D.H. Rouvray and R.B. King, Horwood, Chichester, 2002, pp. 327–337.

I. Lukovits, Constructive enumeration of chiral isomers of alkanes, *Croat. Chem. Acta* 77 (2004) 295–300.

I. Lukovits and I. Gutman, On Morgan trees, *Croat. Chem. Acta* 75 (2002) 563–576.

A. Miličević and N. Trinajstić, Combinatorial enumeration in chemistry, in *Chemical modelling: Applications and theory*, ed. A. Hinchliffe, Vol. 4, Royal Society of Chemistry, Cambridge, 2006, pp. 408–472.

M. Randić, On the recognition of identical graphs representing molecular topology, *J. Chem. Phys.* 60 (1974) 3920–3928.

N. Trinajstić, *Chemical graph theory*, Vols. I and II, CRC, Boca Raton, FL, 1983.

N. Trinajstić, *Chemical graph theory*, 2nd ed., CRC, Boca Raton, FL, 1992.

R.J. Wilson, *Introduction to graph theory*, Oliver and Boyd, Oxford, 1972.

# 2 The Adjacency Matrix and Related Matrices

Adjacency matrices are *square* (and typically *sparse*) $V \times V$ or $E \times E$ symmetric matrices that reflect the adjacencies between vertices or edges in graphs ($V$ = the number of vertices, $E$ = the number of edges). Variants of adjacency matrices, called augmented adjacency matrices (Randić, 1991a), are adjacency matrices that possess nonzero values on the main diagonal (Trinajstić, 1983, 1992; Cvetković et al., 1995). Once the adjacency matrix is known, the related graph can easily be reconstructed. The structure of the adjacency matrix depends, however, on the labeling of a graph. Therefore, the adjacency matrix is *not* a graph invariant. An invariant of a graph $G$ is a number associated with $G$ that has the same value for any graph isomorphic to $G$ (Harary, 1971).

## 2.1  THE VERTEX-ADJACENCY MATRIX OF SIMPLE GRAPHS

The *vertex-adjacency matrix* or binary matrix, denoted by $^v\mathbf{A}$, of a vertex-labeled connected simple graph $G$ with $V$ vertices is a square $V \times V$ matrix, which is determined by the adjacencies of vertices in $G$ (Harary, 1971):

$$\left[\,^v\mathbf{A}\,\right]_{ij} = \begin{cases} 1 & \text{if vertices } i \text{ and } j \text{ are adjacent} \\ 0 & \text{otherwise} \end{cases} \tag{2.1}$$

The term *vertex-adjacency matrix* was first used in chemical graph theory by Mallion in his interesting paper on graph-theoretical aspects of the ring current theory (Mallion, 1975). Below we give the vertex-adjacency matrix of the vertex-labeled graph $G_1$ (see structure $A$ in Figure 2.1).

$$^v\mathbf{A}(G_1) = \begin{bmatrix} 0 & 1 & 0 & 0 & 0 & 0 & 0 \\ 1 & 0 & 1 & 0 & 0 & 0 & 0 \\ 0 & 1 & 0 & 1 & 0 & 1 & 0 \\ 0 & 0 & 1 & 0 & 1 & 0 & 0 \\ 0 & 0 & 0 & 1 & 0 & 1 & 0 \\ 0 & 0 & 1 & 0 & 1 & 0 & 1 \\ 0 & 0 & 0 & 0 & 0 & 1 & 0 \end{bmatrix}$$

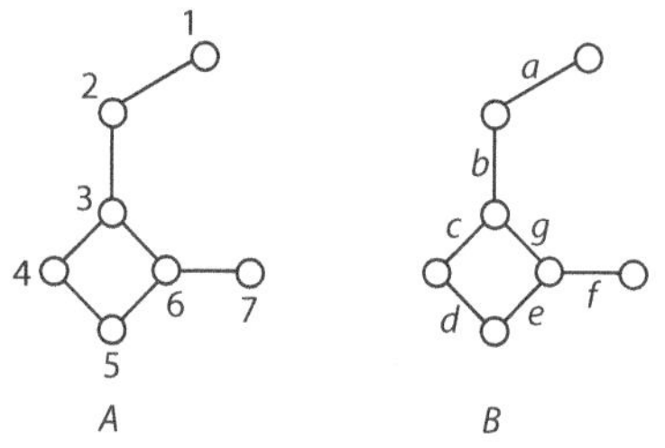

**FIGURE 2.1**    A vertex-labeled (*A*) and edge-labeled (*B*) graph $G_1$.

It is evident that $^v\mathbf{A}$ is a symmetric matrix with a zero diagonal. Therefore, the transpose $^v\mathbf{A}^T$ of the vertex-adjacency matrix leaves $^v\mathbf{A}$ unchanged:

$$^v\mathbf{A}^T = {}^v\mathbf{A} \tag{2.2}$$

The *inverse* of the vertex-adjacency matrix $^v\mathbf{A}^{-1}$ is defined by

$$^v\mathbf{A}^{-1}\,{}^v\mathbf{A} = {}^v\mathbf{A}\,{}^v\mathbf{A}^{-1} = \mathbf{I} \tag{2.3}$$

where $\mathbf{I}$ is the $V \times V$ *unit matrix*. The $^v\mathbf{A}^{-1}$ matrix is defined as

$$^v\mathbf{A}^{-1} = \mathrm{adj}\ {}^v\mathbf{A}/\det{}^v\mathbf{A} \tag{2.4}$$

where adj $^v\mathbf{A}$ is the *matrix adjoint* to $^v\mathbf{A}$. The matrix adjoint to $^v\mathbf{A}$ can be obtained by first replacing each matrix element $[^v\mathbf{A}]_{ij}$ by its cofactor in det $^v\mathbf{A}$, and then transposing rows and columns. If the vertex-adjacency matrix $^v\mathbf{A}$ is to have an inverse $^v\mathbf{A}^{-1}$, the determinant of $^v\mathbf{A}$ must not vanish.

If the vertex-adjacency matrix is associated with the graph $G$ composed of two components $G_a$ and $G_b$,

$$G = G_a \cup G_b \tag{2.5}$$

then $^v\mathbf{A}$ has the block-diagonal form:

$$^v\mathbf{A}(G) = \begin{bmatrix} ^v\mathbf{A}(G_a) & \mathbf{0} \\ \mathbf{0} & ^v\mathbf{A}(G_b) \end{bmatrix}$$

where $\mathbf{0}$ are the *zero matrices* of the size possessed by the components.

A simple yet useful result concerns the vertex-adjacency matrix of bipartite graphs. A *bipartite graph* $G$ is a graph whose vertex-set $V(G)$ can be partitioned into two nonempty subsets $V_1$ and $V_2$ such that every edge in $G$ connects $V_1$ and $V_2$. Therefore, the first neighbors of vertices in $V_1$ are contained in $V_2$, and vice versa. It should be noted that acyclic graphs are always bipartite. A simple theorem due to König (1936) is very helpful for quick determination as to whether a given polycyclic graph is bipartite or not: a polycyclic graph is bipartite if, and only if, all its cycles are *even*-membered.

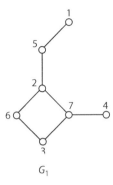

$G_1$

**FIGURE 2.2** Graph $G_1$ with conveniently labeled vertices to give the vertex-adjacency matrix in the block-diagonal form.

In chemistry, bipartite graphs are used to represent *alternant* structures. If a bipartite graph is labeled in such a way that vertices 1, 2, ..., $p$ belong to the subset $V_1$ and vertices $p + 1, p + 2, ..., p + q (= V)$ are in the subset $V_2$, then the corresponding vertex-adjacency matrix is given by

$$^{v}\mathbf{A}(G) = \begin{bmatrix} \mathbf{0} & \mathbf{B} \\ \mathbf{B}^T & \mathbf{0} \end{bmatrix}$$

where $\mathbf{B}$ is a submatrix with dimensions $p \times q$, $\mathbf{B}^T$ is its transpose, and $\mathbf{0}$ are the zero matrices of the size possessed by the submatrices. The consequence of this result is that eigenvalues of the vertex-adjacency matrix of bipartite graphs are paired (Coulson and Rushbroke, 1940). Graph $G_1$ in Figure 1.1 is an example of the bipartite graph. In Figure 2.2, we give the labeling of vertices in $G_1$ resulting from splitting its vertices into subsets $V_1$ and $V_2$, and below it the corresponding block-diagonal form of the vertex-adjacency matrix of $G_1$.

$$^{v}\mathbf{A}(G_1) = \begin{bmatrix} 0 & 0 & 0 & 0 & 1 & 0 & 0 \\ 0 & 0 & 0 & 0 & 1 & 1 & 1 \\ 0 & 0 & 0 & 0 & 0 & 1 & 1 \\ 0 & 0 & 0 & 0 & 0 & 0 & 1 \\ 1 & 1 & 0 & 0 & 0 & 0 & 0 \\ 0 & 1 & 1 & 0 & 0 & 0 & 0 \\ 0 & 1 & 1 & 1 & 0 & 0 & 0 \end{bmatrix}$$

The vertex-adjacency matrix $^{v}\mathbf{A}$, the higher-order vertex-adjacency matrices $^{v}\mathbf{A}_{\kappa}$ ($\kappa = 2, 3, ...$), and the higher-power vertex-adjacency matrices $^{v}\mathbf{A}^{\lambda}$ ($\lambda = 2, 3, ...$) appear to be useful for generating a variety of molecular descriptors (Hosoya, 1971; Trinajstić, 1983, 1992; Barysz et al., 1986; Gutman and Polansky, 1986; Rücker and

Rücker, 1993, 1999, 2003; Devillers and Balaban, 1999; Karelson, 2000; Todeschini and Consonni, 2000, 2009; Gutman et al., 2001; Lukovits et al., 2002; Diudea, 2001; Lukovits and Trinajstić, 2003), but also for other purposes (Marcus, 1963; Randić, 1980; Randić et al., 1983; Rücker and Rücker, 1991, 2003).

The κ-th order vertex-adjacency matrix $^v\mathbf{A}_\kappa$ is defined as

$$
\left[ ^v\mathbf{A}_\kappa \right]_{ij} = \begin{cases} 1 & \text{if the vertex } i \text{ is the } \kappa\text{-th neighbor of the vertex } j \\ 0 & \text{otherwise} \end{cases} \tag{2.6}
$$

For example, the second-order (κ = 2) vertex-adjacency matrix $^v\mathbf{A}_2$ of the vertex-labeled graph $G_1$ (see structure A in Figure 2.1) is

$$
^v\mathbf{A}_2(G_1) = \begin{bmatrix} 0 & 0 & 1 & 0 & 0 & 0 & 0 \\ 0 & 0 & 0 & 1 & 0 & 1 & 0 \\ 1 & 0 & 0 & 0 & 1 & 0 & 1 \\ 0 & 1 & 0 & 0 & 0 & 1 & 0 \\ 0 & 0 & 1 & 0 & 0 & 0 & 1 \\ 0 & 1 & 0 & 1 & 0 & 0 & 0 \\ 0 & 0 & 1 & 0 & 1 & 0 & 0 \end{bmatrix}
$$

The higher-power vertex-adjacency matrices $^v\mathbf{A}^\lambda$ (λ = 2, 3, …) can be obtained by the matrix-multiplication rules. The squared adjacency matrix $^v\mathbf{A}^2$ of the vertex-labeled graph $G_1$ (see structure A in Figure 2.1) is presented below:

$$
^v\mathbf{A}^2(G_1) = \begin{bmatrix} 1 & 0 & 1 & 0 & 0 & 0 & 0 \\ 0 & 2 & 0 & 1 & 0 & 1 & 0 \\ 1 & 0 & 3 & 0 & 2 & 0 & 1 \\ 0 & 1 & 0 & 2 & 0 & 2 & 0 \\ 0 & 0 & 2 & 0 & 2 & 0 & 1 \\ 0 & 1 & 0 & 2 & 0 & 3 & 0 \\ 0 & 0 & 1 & 0 & 1 & 0 & 1 \end{bmatrix}
$$

Walks can be generated from powers of the vertex-adjacency matrix (Hinchliffe, 2004). A *walk* in a graph is an alternating sequence of vertices and edges, such that each edge begins and ends with the vertices immediately preceding and following it (Harary, 1971). The *self-returning walk* is a walk that starts and ends at the same vertex. The length of the walk is the number of edges contained in it. Repetition of vertices or edges is allowed in a walk. The number of walks of length λ beginning at vertex $i$ and ending at vertex $j$ is given by the $i,j$-element of the λ-th power of the vertex-adjacency matrix: $[^v\mathbf{A}^\lambda]_{ij}$. The number of self-returning walks of length λ is

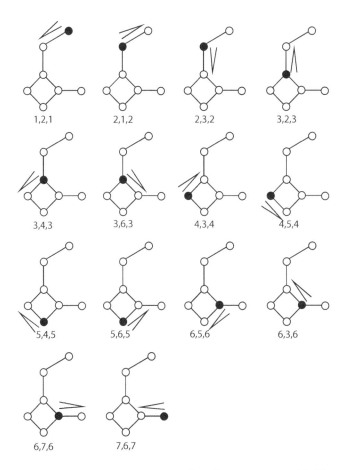

**FIGURE 2.3** All self-returning walks of length 2 on $G_1$ (see structure $A$ in Figure 2.2).

given by the $i,i$-element of the $\lambda$-th power of the vertex-adjacency matrix: $[{}^{v}\mathbf{A}^{\lambda}]_{ii}$. Walks have been extensively applied, for example, as measures of the complexity of graphs, molecules, and surfaces (Nikolić et al., 2000a, 2003b; Rücker and Rücker, 2000, 2001; Randić, 2001a; Randić and Plavšić, 2002, 2003; Lukovits et al., 2002; Konc et al., 2006; Janežić et al., 2009) and for discrimination and ordering of chemical structures (Razinger, 1986). All self-returning walks and all walks of length 2 on $G_1$ are, respectively, illustrated in Figures 2.3 and 2.4.

The *permanent* (also referred to as the positive determinant) of the vertex-adjacency matrix *per* ${}^{v}\mathbf{A}$ can be used to enumerate the number of Kekulé structures $K$, or in the graph-theoretical terminology *1-factors* (Harary, 1971; Cvetković et al., 1995) or *dimers* (Percus, 1969, 1971; Cvetković et al., 1995), of alternant structures (Minc, 1978; Cvetković et al., 1972, 1974a; Kasum et al., 1981; Schultz et al., 1992; Cash, 1995; Torrens, 2002; Jiang et al., 2006):

$$per\ {}^{v}\mathbf{A} = (K)^2 \qquad (2.7)$$

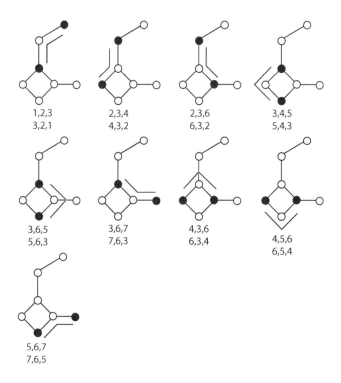

1,2,3    2,3,4    2,3,6    3,4,5
3,2,1    4,3,2    6,3,2    5,4,3

3,6,5    3,6,7    4,3,6    4,5,6
5,6,3    7,6,3    6,3,4    6,5,4

5,6,7
7,6,5

**FIGURE 2.4**    All walks of length 2 on $G_1$ (see structure $A$ in Figure 2.2).

Many graph-theoretical descriptors or topological indices are based on the vertex-adjacency matrix, e.g., the total vertex-adjacency index (Bonchev, 2001), the Narumi simple topological index (Narumi, 1987), the Zagreb indices (Gutman and Trinajstić, 1972; Gutman et al., 1975, 2005; Devillers and Balaban, 1999; Todeschini and Consonni, 2000, 2009; Nikolić et al., 2003a; Hu et al., 2003; Gutman and Das, 2004; Das and Gutman, 2004; Zhou, 2004; Zhou and Gutman, 2005; Janežič et al., 2007; Hanson and Vukičević, 2007; Trinajstić et al., 2010; Das and Trinajstić, 2011), the vertex-connectivity index and its variants (Randić, 1975, 2001b; Kier and Hall, 1976, 1986; Trinajstić, 1983, 1992; Bonchev and Kier, 1992; Estrada, 1995a, 1995b, 1995c; Devillers and Balaban, 1999; Pogliani, 2000, 2002; Todeschini and Consonni, 2000, 2009; Basak et al., 2000; Kezele et al., 2002; Li and Gutman, 2006; Gutman and Furtula, 2008; Zhou and Trinajstić, 2009a, 2009b, 2010a, 2010b; Lučić et al., 2010, 2013), overall connectivity indices (Bonchev, 2001; Bonchev and Trinajstić, 2001), the Gordon-Scantlebury index (Gordon and Scantlebury, 1964; Barysz et al., 1986), the Platt index (Platt, 1947; Barysz et al., 1986), the leading eigenvalue of the vertex-adjacency matrix (Cvetković and Gutman, 1977), or the walk-count indices (Marcus, 1963; Randić, 1980; Randić et al., 1983; Barysz et al. 1986;  Rücker and Rücker, 1993, 1999, 2000, 2001, 2003; Gutman et al., 2001; Lukovits et al., 2002; Lukovits and Trinajstić, 2003; Nikolić et al., 2003b; Hu et al., 2003; Bonchev, 2004).

Pogliani and his coworkers (Garcia-Domenech et al., 2008), besides illustrating the uses of the vertex-connectivity index and variable vertex-connectivity index, also proposed the three interpretations of the vertex-connectivity index: a quantum interpretation, a kinetic interpretation, and a geometric interpretation.

Randić (2012) introduced a novel approach to the problem of protein alignments. It is based on superposition of amino acid adjacent matrices of a pair of proteins, which have been modified to record the sequential order of amino acids.

## 2.2 THE LINEAR REPRESENTATION OF THE VERTEX-ADJACENCY MATRIX OF ACYCLIC STRUCTURES

Lukovits (2000, 2002, 2004) and Lukovits and Gutman (2002) offered an approach by which the vertex-adjacency matrix of an acyclic structure can be replaced by a single number, called the *compressed (vertex-) adjacency matrix code*, denoted by CAM. Here we present, besides the CAM code, the N-tuple code of trees that induces the unique labeling of trees (Aringhieri et al., 1999). A graph is acyclic if it does not contain cycles. A tree is a connected acyclic graph.

The N-tuple code has initially been derived for trees (Read, 1976; von Knop et al., 1985, 1987; Trinajstić, 1990; Trinajstić et al., 1991). It consists of a string of nonnegative integers, each representing the degree of a vertex in a tree or subtree. The degree of a vertex in a (molecular) graph is equal to the number of edges meeting at this vertex. To generate the N-tuple code, one has first to identify the vertices of the highest degree and select among them one that will result in a code that produces lexicographically the largest number. After the initial vertex is located, that vertex and the edges adjacent to it are removed and the subtrees thus produced are examined. This means searching for the largest chain, and if several chains of the same length appear, their codes are derived and combined in such a way that the resulting N-tuple code corresponds to the lexicographically highest number. The steps that can speed up the search of the N-tuple code can be summarized as follows: (1) locate vertices of the highest degree, (2) locate the longest path, (3) backtrack to the last past branching point to visit all vertices in that branch, (4) continue the process until all vertices branching from the longest path have been accounted for, and (5) locate the next longest path and continue the process until all vertices have been recorded. It is important to point out that two nonisomorphic trees cannot have the same N-tuple.

The N-tuple codes are brief—their length is given by V, the number of vertices in a tree. Therefore, they belong to the *linear compact codes* (Randić, 1986; Randić et al., 1988, Nikolić and Trinajstić, 1990, Trinajstić et al., 1991), so named because they use a limited number of digits for linearly encoding a given molecular structure.

Another important property of the N-tuple code is that it induces a unique labeling of vertices in an acyclic graph (Randić, 1986). N-tuple codes also order isomeric acyclic graphs in accordance with their mode of branching. Additionally, the N-tuple approach has been used to develop a very powerful computer program for generation and enumeration of various kinds of (chemical) trees (von Knop et al., 1985; Trinajstić et al., 1991).

In Figure 2.5, we give as an illustrative example a branched tree representing the carbon skeleton of 2,2,3-trimethylhexane and the labels of its vertices produced by the N-tuple code. The N-tuple code of 2,2,3-trimethylhexane is 421100000.

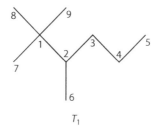

$T_1$

**FIGURE 2.5** A branched tree representing the carbon skeleton of 2,2,3-trimethylhexane and the vertex-labels induced by the N-tuple code.

The vertex-adjacency matrix of $T_1$ is as follows:

$$
{}^{v}\mathbf{A}(T_1) =
\begin{bmatrix}
0 & 1 & 0 & 0 & 0 & 0 & 1 & 1 & 1 \\
1 & 0 & 1 & 0 & 0 & 1 & 0 & 0 & 0 \\
0 & 1 & 0 & 1 & 0 & 0 & 0 & 0 & 0 \\
0 & 0 & 1 & 0 & 1 & 0 & 0 & 0 & 0 \\
0 & 0 & 0 & 1 & 0 & 0 & 0 & 0 & 0 \\
0 & 1 & 0 & 0 & 0 & 0 & 0 & 0 & 0 \\
1 & 0 & 0 & 0 & 0 & 0 & 0 & 0 & 0 \\
1 & 0 & 0 & 0 & 0 & 0 & 0 & 0 & 0 \\
1 & 0 & 0 & 0 & 0 & 0 & 0 & 0 & 0
\end{bmatrix}
$$

The structure of the vertex-adjacency matrix is such that it contains only one non-zero element in each column of its upper (or in each row of its lower) triangle. This fact allows one to replace the vertex-adjacency matrix by the CAM code. Each digit in the CAM code denotes the row in the upper-half of the vertex-adjacency matrix in which the digit 1 is placed. Thus, the CAM code is a linear representation of the vertex-adjacency matrix composed of $V - 1$ entries. Different tree-labelings lead to a different CAM. The resulting CAM code of $T_1$ is 12342111. It should be also noted that the CAM code can be determined directly by inspecting the labeling of a tree.

The vertex-adjacency matrix and the corresponding tree can be easily recovered from the CAM code. Let us consider the following CAM code: 12335567. This code represents the compact form of the following matrix:

$$
{}^{v}\mathbf{A} =
\begin{bmatrix}
0 & 1 & & & & & & & \\
 & 0 & 1 & & & & & & \\
 & & 0 & 1 & 1 & & & & \\
 & & & 0 & & & & & \\
 & & & & 0 & 1 & 1 & & \\
 & & & & & 0 & & 1 & \\
 & & & & & & 0 & & 1 \\
 & & & & & & & 0 & \\
 & & & & & & & & 0
\end{bmatrix}
$$

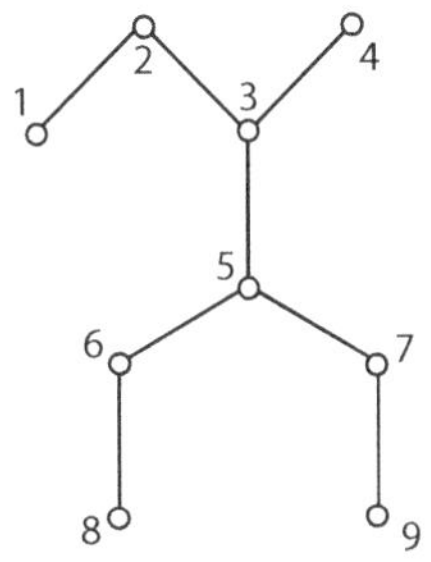

**FIGURE 2.6**   A branched tree representing the carbon skeleton of 3-ethyl-4-methylhexane and the vertex-labels induced by the CAM code.

The corresponding tree is presented in Figure 2.6.

This particular type of CAM code has been introduced by Lukovits (2000), who called it the *lowest degree first* (LDF) code.

## 2.3   LABELING GRAPHS USING THE RANDIĆ PROCEDURE

Randić (1974) proposed a procedure that can be used to check if two graphs are identical or not. He developed an approach for labeling of vertices in graphs that allows the setting up of the unique vertex-adjacency matrices. The Randić approach is based on several useful rules (Randić, 1974):

1. Detect terminal, bridge, and branched vertices in a graph, i.e., vertices of degrees 1, 2, and 3, respectively.
2. Number the vertices in a graph as follows: assign smaller numbers to terminal vertices, intermediate numbers to bridges, and the largest numbers to branched vertices.
3. Connected vertices should be given numbers as different as possible.
4. Between two vertices of the same degree, one connected to a nearest neighbor of higher degree should be assigned a smaller number.

We applied these rules to a tree $T_2$ representing the carbon skeleton of 2,3-dimethyhexane and a graph $G_1$ representing the carbon skeleton of 1-ethyl-2-methylcyclobutane. In Figures 2.7 and 2.8, we give the numbering of vertices in $T_2$ and $G_1$ following the above rules.

The corresponding vertex-adjacency matrix of $T_2$ is given as follows:

$$
{}^{v}\mathbf{A}(T_2) = \begin{bmatrix}
0 & 0 & 0 & 0 & 0 & 0 & 0 & 1 \\
0 & 0 & 0 & 0 & 0 & 0 & 0 & 1 \\
0 & 0 & 0 & 0 & 0 & 0 & 1 & 0 \\
0 & 0 & 0 & 0 & 0 & 1 & 0 & 0 \\
0 & 0 & 0 & 0 & 0 & 1 & 1 & 0 \\
0 & 0 & 0 & 1 & 1 & 0 & 0 & 0 \\
0 & 0 & 1 & 0 & 1 & 0 & 0 & 1 \\
1 & 1 & 0 & 0 & 0 & 0 & 1 & 0
\end{bmatrix}
$$

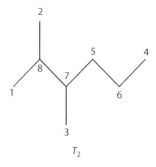

**FIGURE 2.7**  The numbering vertices in $T_2$ following rules 1–4.

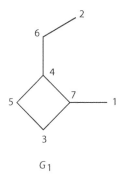

**FIGURE 2.8**  The numbering vertices in $G_1$ following rules 1–4.

The corresponding vertex-adjacency matrix of $G_1$ is given as follows:

$$
{}^{v}\mathbf{A}(G_1) = \begin{bmatrix}
0 & 0 & 0 & 0 & 0 & 0 & 1 \\
0 & 0 & 0 & 0 & 0 & 1 & 0 \\
0 & 0 & 0 & 0 & 1 & 0 & 1 \\
0 & 0 & 0 & 0 & 1 & 1 & 1 \\
0 & 0 & 1 & 1 & 0 & 0 & 0 \\
0 & 1 & 0 & 1 & 0 & 0 & 0 \\
1 & 0 & 1 & 1 & 0 & 0 & 0
\end{bmatrix}
$$

## 2.4  THE VERTEX-ADJACENCY MATRIX OF MULTIPLE GRAPHS

The vertex-adjacency matrix of a vertex-labeled multiple graph $G$ is a square $V \times V$ matrix defined as

$$
\left[{}^{v}\mathbf{A}\right]_{ij} = \begin{cases} m_{ij} & \text{if vertices } i \text{ and } j \text{ are adjacent} \\ 0 & \text{otherwise} \end{cases} \tag{2.8}
$$

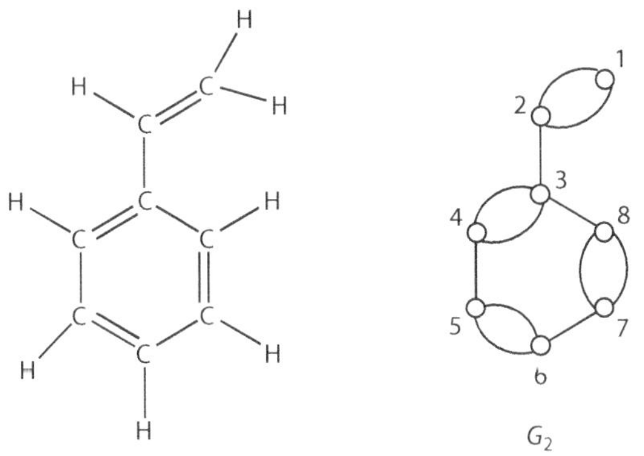

**FIGURE 2.9** The vertex-labeled multiple graph $G_2$ representing the carbon skeleton of one Kekulé structure of styrene.

where $m_{ij}$ is the multiplicity of the $i$-$j$ edge. In Figure 2.9 is given the multiple graph $G_2$ depicting carbon skeleton of one Kekulé structure of styrene. The corresponding vertex-adjacent matrix is given below the figure.

$$
{}^{v}\mathbf{A}(G_2) = \begin{bmatrix}
0 & 2 & 0 & 0 & 0 & 0 & 0 & 0 \\
2 & 0 & 1 & 0 & 0 & 0 & 0 & 0 \\
0 & 1 & 0 & 2 & 0 & 0 & 0 & 1 \\
0 & 0 & 2 & 0 & 1 & 0 & 0 & 0 \\
0 & 0 & 0 & 1 & 0 & 2 & 0 & 0 \\
0 & 0 & 0 & 0 & 2 & 0 & 1 & 0 \\
0 & 0 & 0 & 0 & 0 & 1 & 0 & 2 \\
0 & 0 & 1 & 0 & 0 & 0 & 2 & 0
\end{bmatrix}
$$

## 2.5  THE ATOM-CONNECTIVITY MATRIX

The *atom-connectivity matrix*, denoted by **ACM**, has been proposed by Spialter (1963, 1964a, 1964b) for use in computer-oriented chemical nomenclature. This matrix represents the structural formula of a molecule and is given by

$$
[\mathbf{ACM}]_{ij} = \begin{cases} b_{ij} & \text{if vertices } i \text{ and } j \text{ are adjacent} \\ s_i & \text{if } i = j \\ 0 & \text{otherwise} \end{cases} \tag{2.9}
$$

where $b_{ij}$ is the bond order between atoms $i$ and $j$, and $s_i$ stands for the chemical symbol of the atom $i$. The values of bond orders in most cases are 1, 1.5, 2, and 3 for single, aromatic, double, and triple bonds, respectively. Below we give an example of

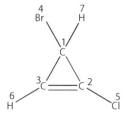

**FIGURE 2.10**  Structural formula of 1-bromo-2-chlorocycloprop-2-ene.

the atom-connectivity matrix of 1-bromo-2-chlorocycloprop-2-ene, whose structural formula is presented in Figure 2.10.

$$
\mathbf{ACM} = \begin{bmatrix}
C & 1 & 1 & 1 & 0 & 0 & 1 \\
1 & C & 2 & 0 & 1 & 0 & 0 \\
1 & 2 & C & 0 & 0 & 1 & 0 \\
1 & 0 & 0 & Br & 0 & 0 & 0 \\
0 & 1 & 0 & 0 & Cl & 0 & 0 \\
0 & 0 & 1 & 0 & 0 & H & 0 \\
1 & 0 & 0 & 0 & 0 & 0 & H
\end{bmatrix}
$$

If only the molecular skeleton without hydrogen atoms is considered, then one gets the hydrogen-suppressed structure. Spialter called the corresponding structural matrix the *hydrogen-suppressed atom-connectivity matrix*, denoted by **HS-ACM** (Spialter, 1964a; von Knop et al., 1975). The advantage of using **HS-ACM** instead of **ACM** at that time (1964) was in reducing the size of the matrix to save computer time.

The **HS-ACM** matrix for 1-bromo-2-chlorocycloprop-2-ene is as follows:

$$
\mathbf{HS-ACM} = \begin{bmatrix}
C & 1 & 1 & 1 & 0 \\
1 & C & 2 & 0 & 1 \\
1 & 2 & C & 0 & 0 \\
1 & 0 & 0 & Br & 0 \\
0 & 1 & 0 & 0 & Cl
\end{bmatrix}
$$

It should also be pointed out that the roots of the atom-connectivity matrix concept go back at least to Balandin (1940), though there are indications that this concept is even older (Randić and Trinajstić, 1994). Balandin constructed *property matrices* by using the symbols of atoms making up the molecule as diagonal elements and molecular properties such as the interatomic distances (in Å), bond dissociation energies (in kcal), and vibrational force constants (in cm$^{-1}$ × 10$^4$) as off-diagonal elements.

The important step in the development of the concept was made by Wheland. In 1946, Wheland (1946) represented molecules as tableaux (connection tables). This Wheland contribution may be regarded as a starting point of the computer-oriented chemical documentation (Trinajstić and Toth, 1986). The Wheland tableau for 1-bromo-2-chlorocycloprop-2-ene is given below and is practically identical to the **ACM** matrix of Spialter.

|      | C | C | C | Br | Cl | H | H |
|------|---|---|---|----|----|---|---|
| C    | — | 1 | 1 | 1  | 0  | 0 | 1 |
| C    | 1 | — | 2 | 0  | 1  | 0 | 0 |
| C    | 1 | 2 | — | 0  | 0  | 1 | 0 |
| Br   | 1 | 0 | 0 | —  | 0  | 0 | 0 |
| Cl   | 0 | 1 | 0 | 0  | —  | 0 | 0 |
| H    | 0 | 0 | 1 | 0  | 0  | — | 0 |
| H    | 1 | 0 | 0 | 0  | 0  | 0 | — |

Since Wheland did not pursue his idea, this result of his remained practically unknown. Thus, for example, Spialter did not mention the Wheland tableaux in his papers. After Wheland, several authors, located either in industry or at the former National Bureau of Standards (now the National Institute for Standards and Technology) in Washington, D.C., developed methods for computer storage and retrieval of chemical structures (Ray and Kirsch, 1957; Gluck, 1965; Meyer, 1969). Spialter himself was at the Wright-Patterson Air Force Base in Fairborn near Dayton, Ohio. Apparently the need for computer-based systems for chemical documentation was perceived much earlier in applied research than in academic research.

## 2.6 THE BOND-ELECTRON MATRIX

The bond-electron matrices **BE** of Ugi and coworkers (Dugundji and Ugi, 1973; Ugi et al., 1979, 1990) represent a variant of the atom-bond matrices. They are defined as

$$
[\mathbf{BE}]_{ij} = \begin{cases} b_{ij} & \text{if vertices } i \text{ and } j \text{ are adjacent} \\ b_{ii} & \text{if } i = j \\ 0 & \text{otherwise} \end{cases} \tag{2.10}
$$

where $b_{ii}$, the $i$-th diagonal entry, stands for the number of free valence electrons (electrons not participating in the bonding) and $b_{ij}$, the off-diagonal entry between the $i$-th row and $j$-th column, represents the formal bond order between atoms $i$ and $j$, respectively.

The **BE** matrix for 1-bromo-2-chlorocycloprop-2-ene is related to its **ACM** matrix:

$$\mathbf{BE} = \begin{bmatrix} 0 & 1 & 1 & 1 & 0 & 0 & 1 \\ 1 & 0 & 2 & 0 & 1 & 0 & 0 \\ 1 & 2 & 0 & 0 & 0 & 1 & 0 \\ 1 & 0 & 0 & 6 & 0 & 0 & 0 \\ 0 & 1 & 0 & 0 & 6 & 0 & 0 \\ 0 & 0 & 1 & 0 & 0 & 0 & 0 \\ 1 & 0 & 0 & 0 & 0 & 0 & 0 \end{bmatrix}$$

It should be noted that the sum of the elements in the $i$-th row or column of a **BE** matrix gives the number of valence electrons $n_i$ of atom $i$:

$$n_i = \sum_{j=1}^{N} b_{ij} \tag{2.11}$$

where $N$ is the size of the **BE** matrix.

Ugi and coworkers (Dugundji and Ugi, 1973; Ugi et al., 1979, 1990) also studied the conversion of a molecule $A_1$ into its isomeric counterpart $A_2$. This can be schematized as

$$A_1 \rightarrow A_2 \tag{2.12}$$

This transformation can be described by means of the following equation:

$$\mathbf{BE}(A_1) + \mathbf{RM} = \mathbf{BE}(A_2) \tag{2.13}$$

In (2.13), the symbols $\mathbf{BE}(A_1)$ and $\mathbf{BE}(A_2)$ stand for the bond-electron matrices of reactants $A_1$ and products $A_2$, and $\mathbf{RM}$ symbolizes the reaction matrix, respectively.

The off-diagonal entries of the reaction matrix $(\mathbf{RM})_{ij}$,

$$(\mathbf{RM})_{ij} = [\mathbf{BE}(A_1)]_{ij} - [\mathbf{BE}(A_2)]_{ij} \tag{2.14}$$

denote the changes of the formal order of bonds between $A_1$ and $A_2$, whereas the diagonal entries $(\mathbf{RM})_{ii}$ denote the change in the number of the valence electrons at the atoms of the molecule $A_2$. Thus, the matrix $\mathbf{RM}$ may be regarded as a scheme of electron redistribution. This is exemplified using the Dugundji-Ugi transformation of hydrogen cyanide (HCN) into hydrogen isocyanide (HNC):

$$H - C \equiv N: \rightarrow H - \overset{+}{N} = \overset{-}{C}$$

The corresponding Equation (2.13) is given by

$$\begin{bmatrix} 0 & 1 & 0 \\ 1 & 0 & 3 \\ 0 & 3 & 2 \end{bmatrix} + \begin{bmatrix} 0 & -1 & 1 \\ -1 & 2 & 0 \\ 1 & 0 & -2 \end{bmatrix} = \begin{bmatrix} 0 & 0 & 1 \\ 0 & 2 & 3 \\ 1 & 3 & 0 \end{bmatrix}$$

Dugundji, Ugi, and his coworkers (Dugundji and Ugi, 1973; Ugi et al., 1979, 1990) used the bond-electron matrices **BE** and the reaction matrices **RM** as a basis for computer programs for the deductive solution of a variety of chemical problems, such as the classification and documentation of structures, substructures, and reactions, the prognosis of reaction products, the design of synthetic routes, the construction of networks of mechanistic and preparative pathways, the prediction of chemical reactions, etc.

## 2.7 THE EDGE-ADJACENCY MATRIX

The *edge-adjacency matrix*, denoted by $^e\mathbf{A}$, of an edge-labeled connected graph $G$ is a square $E \times E$ matrix, which is determined by the adjacencies of edges (Rouvray, 1976; Trinajstić, 1983, 1992):

$$\left[{}^e\mathbf{A}\right]_{ij} = \begin{cases} 1 & \text{if edges } i \text{ and } j \text{ are adjacent} \\ 0 & \text{otherwise} \end{cases} \tag{2.15}$$

Below we give the edge-adjacency matrix of the edge-labeled graph $G_1$ (see structure $B$ in Figure 2.1).

$${}^e\mathbf{A}(G_1) = \begin{bmatrix} 0 & 1 & 0 & 0 & 0 & 0 & 0 \\ 1 & 0 & 1 & 0 & 0 & 0 & 1 \\ 0 & 1 & 0 & 1 & 0 & 0 & 1 \\ 0 & 0 & 1 & 0 & 1 & 0 & 0 \\ 0 & 0 & 0 & 1 & 0 & 1 & 1 \\ 0 & 0 & 0 & 0 & 1 & 0 & 1 \\ 0 & 1 & 1 & 0 & 1 & 1 & 0 \end{bmatrix}$$

It should be noted that the vertex-adjacency matrix *uniquely* determines a graph, but the edge-adjacency matrix *does not*; that is, there are known nonisomorphic graphs with *identical* edge-adjacency matrices. For example, a pair of nonisomorphic graphs the three-point star $S_3$ and the cycle on three vertices $C_3$ possessing identical edge-adjacency matrices is given in Figure 2.11.

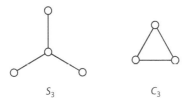

$S_3$ $C_3$

**FIGURE 2.11** A pair of nonisomorphic graphs consisting of the 3-star $S_3$ and the 3-cycle $C_3$ that possess the same edge-adjacency matrix.

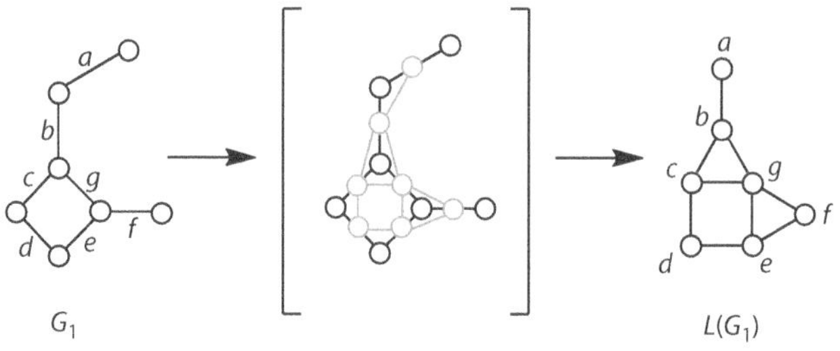

**FIGURE 2.12**   Construction of the line graph $L(G_1)$ from the graph $G_1$.

Below we give the edge-adjacency matrix that represents both $S_3$ and $C_3$:

$$^e\mathbf{A}(S_3) = {}^e\mathbf{A}(C_3) = \begin{bmatrix} 0 & 1 & 1 \\ 1 & 0 & 1 \\ 1 & 1 & 0 \end{bmatrix}$$

$S_3$ and $C_3$ clearly possess *different* vertex-adjacency matrices—they are graphs of different sizes—and they possess different numbers of vertices.

It should also be noted that the edge-adjacency matrix of a graph $G$ is identical to the vertex-adjacency matrix of the corresponding *line graph* $L(G)$ of $G$:

$$^e\mathbf{A}(G) = {}^v\mathbf{A}(L(G)) \tag{2.16}$$

This must be so because the edges in $G$ are replaced by vertices in $L(G)$ (Harary, 1971). In Figure 2.12, we show the construction of the line graph $L(G_1)$ from graph $G_1$.

It is interesting to note that the line graph concept, though not in the explicit mathematical formalism, may be traced back to the beginnings of structural chemistry. Thus, van't Hoff represented simple organic molecules in terms of points and lines, that is, in terms of the line graphs of the modern structural formulas (e.g., Gutman et al., 1998).

A number of the topological indices mentioned above can be reformulated in terms of the edge-degrees instead of the vertex-degrees, e.g., the total edge-adjacency index (Todeschini and Consonni, 2000, 2009), the reformulated Zagreb indices (Milićević et al., 2004; Peng et al., 2004), the edge-connectivity index (Estrada, 1995a, 1995b; Gutman and Estrada, 1996; Trinajstić et al., 1997; Nikolić et al., 1998, 1999a; Basak et al., 2000), the reformulated Gordon-Scantlebury index (Todeschini and Consonni, 2000, 2009), and the reformulated Platt index (Todeschini and Consonni, 2000, 2009).

## 2.8   THE VERTEX-ADJACENCY MATRIX OF WEIGHTED GRAPHS

Weighted graphs in chemistry usually represent heterosystems (Mallion et al., 1974a, 1974b, 1975; Graovac and Trinajstić, 1975a; Graovac, et al., 1975). Molecules containing heteroatoms and heterobonds are represented by the vertex- and edge-weighted graphs (Trinajstić, 1983, 1992). A vertex- and edge-weighted graph $G_{vew}$ is a graph that has one or more of its vertices and edges distinguished in some

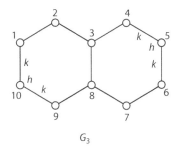

$G_3$

**FIGURE 2.13** A vertex- and edge-weighted graph $G_3$ representing the carbon skeleton 2,6-diazanaphthalene.

way from other vertices and edges in $G_{vew}$. These different vertices and edges are weighted—their weights are usually identified by parameters $h$ and $k$, respectively. In Figure 2.13, we give a vertex- and edge-weighted graph $G_3$ corresponding, for example, to 2,6-diazanaphthalene.

The vertex-adjacency matrix of the vertex- and edge-weighted graph $^v\mathbf{A}(G_{vew})$ is defined by

$$
\left[{}^v\mathbf{A}(G_{vew})\right]_{ij} = \begin{cases} k & \text{if the edge } i-j \text{ is weighted} \\ 1 & \text{if the edge } i-j \text{ is not weighted} \\ h & \text{if } i=j \text{ and if the vertex } i \text{ is weighted} \\ 0 & \text{otherwise} \end{cases} \tag{2.17}
$$

The parameters $h$ and $k$ depend, respectively, on the chemical nature of the corresponding atoms and bonds in a molecule. Some people select for them the values of the Hückel parameters for heteroatoms and heterobonds. Below is given the vertex-adjacency matrix for $G_3$ from Figure 2.13.

$$
^v\mathbf{A}(G_3) = \begin{bmatrix}
0 & 1 & 0 & 0 & 0 & 0 & 0 & 0 & 0 & k \\
1 & 0 & 1 & 0 & 0 & 0 & 0 & 0 & 0 & 0 \\
0 & 1 & 0 & 1 & 0 & 0 & 0 & 1 & 0 & 0 \\
0 & 0 & 1 & 0 & k & 0 & 0 & 0 & 0 & 0 \\
0 & 0 & 0 & k & h & k & 0 & 0 & 0 & 0 \\
0 & 0 & 0 & 0 & k & 0 & 1 & 0 & 0 & 0 \\
0 & 0 & 0 & 0 & 0 & 1 & 0 & 1 & 0 & 0 \\
0 & 0 & 1 & 0 & 0 & 0 & 1 & 0 & 1 & 0 \\
0 & 0 & 0 & 0 & 0 & 0 & 0 & 1 & 0 & k \\
k & 0 & 0 & 0 & 0 & 0 & 0 & 0 & k & h
\end{bmatrix}
$$

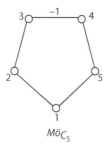

$M\ddot{o}C_5$

**FIGURE 2.14**    Labeled five-membered Möbius cycle.

## 2.9    THE VERTEX-ADJACENCY MATRIX OF MÖBIUS GRAPHS

*Möbius graphs* are a kind of edge-weighted graphs in which at least one edge-weight is –1 (Graovac and Trinajstić, 1975a, 1975b, 1976; Gutman, 1978). They are used to represent the Möbius systems (Heilbronner, 1964; Herges, 2006). The vertex-adjacency matrix of Möbius graphs is a symmetric $V \times V$ matrix:

$$\left[ {}^{v}\mathbf{A} \right]_{ij} = \begin{cases} 1 & \text{if the edge } i - j \text{ is positively weighted} \\ -1 & \text{if the edge } i - j \text{ is negatively weighted} \\ 0 & \text{otherwise} \end{cases} \tag{2.18}$$

In Figure 2.14, we give the five-membered Möbius cycle, denoted by ${}^{M\ddot{o}}C_5$. The corresponding vertex-adjacency matrix is given below the figure.

$${}^{v}\mathbf{A}\left( {}^{M\ddot{o}}C_5 \right) = \begin{bmatrix} 0 & 1 & 0 & 0 & 1 \\ 1 & 0 & 1 & 0 & 0 \\ 0 & 1 & 0 & -1 & 0 \\ 0 & 0 & -1 & 0 & 1 \\ 1 & 0 & 0 & 1 & 0 \end{bmatrix}$$

Möbius graphs are also known as *signed graphs* (e.g., Roberts, 1979; Balasubramanian, 1989). These graphs found use in the Hückel-Möbius rules (e.g., Klein and Trinajstić, 1984).

## 2.10    THE AUGMENTED VERTEX-ADJACENCY MATRIX

Randić introduced in 1991 the variable vertex-connectivity index (Randić, 1991a) using the concept of the *augmented vertex-adjacency matrix* (Randić, 1991a, 1991b; Randić et al., 2001). Variants of adjacency matrices, called the augmented vertex-adjacency matrices and denoted by ${}^{av}\mathbf{A}$, are the vertex-adjacency matrices that possess nonzero values on the main diagonal:

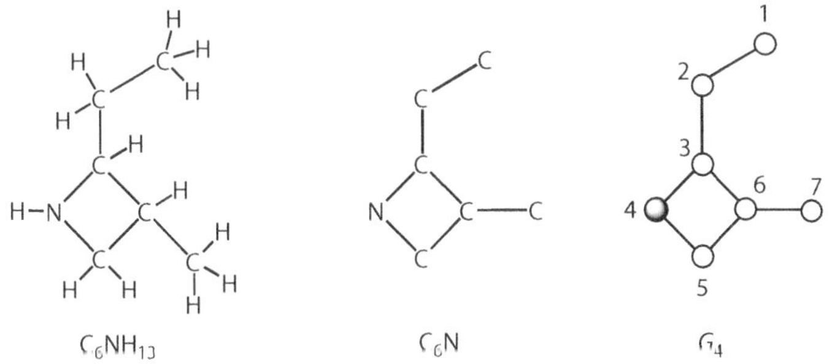

**FIGURE 2.15** A vertex-labeled vertex-weighted graph $G_4$ representing the carbon skeleton $C_6N$ of 2-ethyl-3-methyl-1-azacyclobutane $C_6NH_{13}$. The weighted vertex is denoted by the dark circle.

$$\left[ {}^{av}\mathbf{A} \right]_{ij} = \begin{cases} 1 & \text{if vertices } i \text{ and } j \text{ are adjacent} \\ w_{ii} & \text{if } i = j \\ 0 & \text{otherwise} \end{cases} \quad (2.19)$$

where $w_{ii}$ is the weight at the vertex $i$.

Augmented vertex-adjacency matrices were introduced to be used for *vertex-weighted* graphs, that is, graphs with one or more of their vertices distinguished in some way from the rest of their vertices (Trinajstić, 1983, 1992). But the augmented vertex-adjacency matrices are also used via appropriate molecular descriptors to assess structural differences, such as, for example, the relative role of carbon atoms of acyclic and cyclic parts in alkylcycloalkanes (Randić et al., 2001).

An example of a vertex-weighted graph is shown in Figure 2.15. Below the figure, we give the augmented vertex-adjacency matrix of $G_4$, in which the set of vertices is split in two subsets: the first containing six vertices of one kind (denoted by $x$), and the second containing one vertex of a different kind (denoted by $y$).

$$ {}^{av}\mathbf{A}(G_4) = \begin{bmatrix} x & 1 & 0 & 0 & 0 & 0 & 0 \\ 1 & x & 1 & 0 & 0 & 0 & 0 \\ 0 & 1 & x & 1 & 0 & 1 & 0 \\ 0 & 0 & 1 & y & 1 & 0 & 0 \\ 0 & 0 & 0 & 1 & x & 1 & 0 \\ 0 & 0 & 1 & 0 & 1 & x & 1 \\ 0 & 0 & 0 & 0 & 0 & 1 & x \end{bmatrix} $$

Several topological indices have been used in their variable form, e.g., two formulations of the variable Zagreb indices (Miličević and Nikolić, 2004; Nikolić et al., 2004; Miličević et al., 2004a) and the variable vertex-connectivity index (Randić, 1991a, 1991b; Todeschini and Consonni, 2000, 2009; Randić and Basak, 2001; Yang and Zhong, 2003; Randić et al., 2001, 2004). The variable Zagreb indices and the

variable connectivity indices represent generalizations of the original indices introduced four decades ago (Gutman and Trinajstić, 1972; Gutman et al., 1975; Randić,
1975). They found use in structure-property-activity modeling (Lučić et al., 1999;
Todeschini and Consonni, 2000, 2009; Piližota et al., 2004).

## 2.11   THE EDGE-WEIGHTED EDGE-ADJACENCY MATRIX

The *edge-weighted edge-adjacency matrix*, denoted by $^{ew}\mathbf{A}$, has been introduced by
Estrada (1995a, 1995b). It is a square unsymmetric $E \times E$ matrix defined as

$$\left[ ^{ew}\mathbf{A} \right]_{ij} = \begin{cases} 1 & \text{if edges } i \text{ and } j \text{ are adjacent} \\ k & \text{if the edge } i - j \text{ is weighted} \\ 0 & \text{otherwise} \end{cases} \quad (2.20)$$

As an example of the edge-weighted edge-adjacency matrix, we present this matrix
for the edge-weighted graph $G_5$ depicted in Figure 2.16.

The $^{ew}\mathbf{A}$ matrix of $G_5$ is as follows:

$$^{ew}\mathbf{A}(G_5) = \begin{bmatrix} 0 & 1 & 0 & 0 & 0 & 0 & 0 \\ 1 & 0 & k & 0 & 0 & 0 & 1 \\ 0 & k & 0 & k^2 & 0 & 0 & k \\ 0 & 0 & k^2 & 0 & k & 0 & 0 \\ 0 & 0 & 0 & k & 0 & 1 & 1 \\ 0 & 0 & 0 & 0 & 1 & 0 & 1 \\ 0 & 1 & k & 0 & 1 & 1 & 0 \end{bmatrix}$$

The parameter $k$ is the edge-parameter (the bond-parameter), and Estrada presented in his papers (Estrada, 1995a, 1999) the values of this parameter for most of
the bonds common in organic compounds. Estrada used the edge-connectivity index
to predict successfully the molecular volumes of 112 aliphatic organic compounds,
including aldehydes, ketones, ethers, thioethers, tertiary amines, and alyl halides.

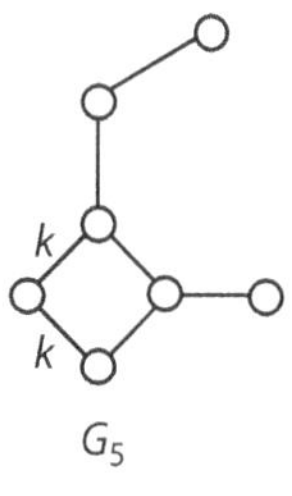

$G_5$

**FIGURE 2.16**   Edge-weighted graph $G_5$.

## 2.12   THE BURDEN MATRIX

The *Burden matrix* (Burden, 1989; Todeschini and Consonni, 2000, 2009; Consonni and Todeschini, 2012) is a special vertex- and edge-weighted adjacency matrix, denoted by **BM**. It is defined as

$$[\mathbf{BM}]_{ij} = \begin{cases} b_{ij}/10 & \text{if vertices } i \text{ and } j \text{ are adjacent} \\ Z_i & \text{if } i = j \\ 1/1000 & \text{otherwise} \end{cases} \tag{2.21}$$

where the diagonal elements $[\mathbf{BM}]_{ii}$ are the atomic numbers $Z_i$ of atoms $i$, while the off-diagonal elements $[\mathbf{BM}]_{ij}$, representing the connected atoms, are given weights $b_{ij}/10 =$ 0.100, 0.200, 0.300, and 0.150 for single, double, triple, and aromatic bonds, respectively. Note $b_{ij}$ is the conventional bond order, being equal to 1, 2, 3, and 1.5 for single, double, triple, and aromatic bonds, respectively. The off-diagonal matrix elements, representing nonconnected atoms, are given weights 0.001. The matrix elements corresponding to terminal atoms are given additional weights 0.010. As an example, we give below the Burden matrix for a vertex-labeled vertex-weighted graph $G_4$ representing the carbon skeleton of 2-ethyl-3-methyl-1-azacyclobutane (see Figure 2.15).

$$\mathbf{BM}(G_4) = \begin{bmatrix} 6 & 0.110 & 0.001 & 0.001 & 0.001 & 0.001 & 0.001 \\ 0.110 & 6 & 0.100 & 0.001 & 0.001 & 0.001 & 0.001 \\ 0.001 & 0.100 & 6 & 0.100 & 0.001 & 0.100 & 0.001 \\ 0.001 & 0.001 & 0.100 & 7 & 0.100 & 0.001 & 0.001 \\ 0.001 & 0.001 & 0.001 & 0.100 & 6 & 0.100 & 0.001 \\ 0.001 & 0.001 & 0.100 & 0.001 & 0.100 & 6 & 0.110 \\ 0.001 & 0.001 & 0.001 & 0.001 & 0.001 & 0.110 & 6 \end{bmatrix}$$

Several generalizations of the Burden matrix have been proposed by, e.g., Ivanciuc (2001) and Sheridan (2002). Ivanciuc generalized the definition of the Burden matrix by replacing the atomic numbers $Z_i$ with a vertex weighting scheme $W_i$:

$$\left[{}^I\mathbf{BM}\right]_{ij} = \begin{cases} b_{ij}/10 & \text{if vertices } i \text{ and } j \text{ are adjacent} \\ W_i & \text{if } i = j \\ 0.001 & \text{otherwise} \end{cases} \tag{2.22}$$

Sheridan modified the Burden matrix in the following way:

$$\left[{}^S\mathbf{BM}\right]_{ij} = \begin{cases} 4/d_{ij} & \text{if vertices } i \text{ and } j \text{ are adjacent} \\ Z_i + 0.1\,\delta_i + 0.01\,n_i^\pi & \text{if } i = j \\ 0.001 & \text{otherwise} \end{cases} \tag{2.23}$$

where $d_{ij}$ is the graph-theoretical distance between the $i$-th and $j$-th atoms, $Z_i$ is as before the atomic number of atom $i$, $\delta_i$ is the number of nonhydrogen neighbors of the $i$-th atom, that is, the degree of the vertex $i$, and $n_i^\pi$ is the number of $\pi$-electrons.

The Burden matrix can be used to produce the unique numbering for atoms in a molecule and for substructure searches. The Sheridan modification of the Burden matrix appears to be useful in the design of drug-like compounds (Todeschini and Consonni, 2000, 2009).

## 2.13   THE VERTEX-CONNECTIVITY MATRIX

The *vertex-connectivity matrix*, denoted by $^v\chi$, was introduced by Randić (1992). This matrix is also named the (connectivity) $\chi$ *matrix* (Todeschini and Consonni, 2000, 2009) and the *product-connectivity matrix* (Zhou and Trinajstić, 2010a; Lučić et al., 2014). It can be regarded as an edge-weighted matrix of a graph that is defined as

$$\left[\,^v\chi\,\right]_{ij} = \begin{cases} [d(i)d(j)]^{-1/2} & \text{if vertices } i \text{ and } j \text{ are adjacent} \\ 0 & \text{otherwise} \end{cases} \tag{2.24}$$

where $d(i)$ and $d(j)$ are the degrees of vertices $i$ and $j$. For example, the degrees of vertices in $G_1$ are given in Figure 2.17.

The vertex-connectivity matrix of $G_1$ (using the vertex-labels presented in structure $A$ in Figure 2.1 and vertex-degrees from Figure 2.17) is given below:

$$^v\chi(G_1) = \begin{bmatrix} 0 & 0.707 & 0 & 0 & 0 & 0 & 0 \\ 0.707 & 0 & 0.408 & 0 & 0 & 0 & 0 \\ 0 & 0.408 & 0 & 0.408 & 0 & 0.333 & 0 \\ 0 & 0 & 0.408 & 0 & 0.500 & 0 & 0 \\ 0 & 0 & 0 & 0.500 & 0 & 0.408 & 0 \\ 0 & 0 & 0.333 & 0 & 0.408 & 0 & 0.577 \\ 0 & 0 & 0 & 0 & 0 & 0.577 & 0 \end{bmatrix}$$

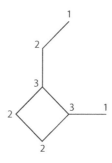

**FIGURE 2.17**   The vertex-degrees in $G_1$.

The summation of elements in the upper (or lower) matrix-triangle gives the vertex-connectivity index of $G_1$ (Gutman, 2002; Hu et al., 2003).

The vertex-connectivity matrix has also been used in computing the connectivity identification (ID) number (Randić, 1984a; Szymanski et al., 1985). The connectivity ID number was successfully tested in QSAR (Randić, 1984b; Carter et al., 1987).

## 2.14  THE EDGE-CONNECTIVITY MATRIX

The *edge-connectivity matrix*, denoted by $^e\chi$, of a graph $G$ is the vertex-connectivity matrix of the corresponding line graph $L(G)$ (Basak et al., 2000; Nikolić et al., 2001; Lučić et al., 2014). As an example, we give $^e\chi$ of $G_1$, which is equal to $^v\chi$ of $L(G_1)$ from Figure 2.12. The edge-degrees of $G_1$ and the vertex-degrees of $L(G_1)$ are shown in Figure 2.18. The degree of an edge is equal to the number of adjacent edges.

$$
^e\chi(G_1) = {}^v\chi[L(G_1)] =
\begin{bmatrix}
0 & 0.577 & 0 & 0 & 0 & 0 & 0 \\
0.577 & 0 & 0.333 & 0 & 0 & 0.289 & 0 \\
0 & 0.333 & 0 & 0.408 & 0 & 0.289 & 0 \\
0 & 0 & 0.408 & 0 & 0.408 & 0 & 0 \\
0 & 0 & 0 & 0.408 & 0 & 0.289 & 0408 \\
0 & 0.289 & 0.289 & 0 & 0.289 & 0 & 0.354 \\
0 & 0 & 0 & 0 & 0.408 & 0.354 & 0
\end{bmatrix}
$$

Summation of the elements in the upper (or lower) matrix-triangle gives the edge-connectivity index of $G_1$ (Estrada, 1995a). It has been applied to structure-property modeling of various classes of molecules (e.g., Estrada, 1995a, 1995b, 1995c; Nikolić et al., 1998).

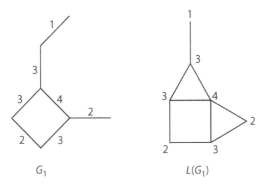

**FIGURE 2.18**   The edge-degrees in $G_1$ and vertex-degrees in $L(G_1)$.

## 2.15  THE SUM-VERTEX-CONNECTIVITY MATRIX

The *sum-vertex-connectivity matrix*, denoted by $^vS$, was introduced independently by Zhou and Trinajstić (2009, 2010a) and Randić et al. (2010). Randić et al. (2010) named this matrix the *distance-weighted adjacency matrix*. It is defined as follows:

$$\left[ ^{v}S \right]_{ij} = \begin{cases} [d(i) + d(j)]^{-1/2} & \text{if vertices } i \text{ and } j \text{ are adjacent} \\ 0 & \text{otherwise} \end{cases} \tag{2.25}$$

The sum-vertex-connectivity matrix of $G_1$ (using the vertex-labels presented in structure $A$ in Figure 2.1 and vertex-degrees from Figure 2.17) is given below:

$$^{v}S(G_1) = \begin{bmatrix} 0 & 0.577 & 0 & 0 & 0 & 0 & 0 \\ 0.577 & 0 & 0.447 & 0 & 0 & 0 & 0 \\ 0 & 0.447 & 0 & 0.447 & 0 & 0.408 & 0 \\ 0 & 0 & 0.447 & 0 & 0.500 & 0 & 0 \\ 0 & 0 & 0 & 0.500 & 0 & 0.447 & 0 \\ 0 & 0 & 0.408 & 0 & 0.447 & 0 & 0.500 \\ 0 & 0 & 0 & 0 & 0 & 0.500 & 0 \end{bmatrix}$$

The summation of matrix elements in the upper or lower triangle gives the sum-vertex-connectivity index. The mathematical properties of this molecular descriptor and its uses in the structure-property modeling are reported in several publications (Todeschini and Consonni, 2009; Zhou and Trinajstić, 2009, 2010a, 2010b, 2010c, 2010d, 2012; Lučić et al., 2009, 2010, 2013, 2014; Randić et al., 2010; Du et al., 2010, 2011; Xing et al., 2010; Vukičević and Trinajstić, 2010; Wang et al., 2011; Nikolić et al., 2012). The structure-property-activity models based on the sum-vertex-connectivity indices parallel those based on the product-vertex-connectivity indices (e.g., Lučić et al., 2010; Nikolić et al., 2012).

## 2.16  THE SUM-EDGE-CONNECTIVITY MATRIX

The *sum-edge-connectivity matrix*, denoted by $^eS$, of a graph $G$ is the sum-vertex-connectivity matrix of the corresponding line graph $L(G)$ (Lučić et al., 2014):

$$^{e}S(G) = ^{v}S[L(G)] \tag{2.26}$$

For example, the sum-edge-connectivity matrix of the graph $G_1$ is equal to the sum-vertex-connectivity matrix of the corresponding line graph $L(G_1)$:

$$
{}^e\mathbf{S}(G_1) = {}^v\mathbf{S}[L(G_1)] = \begin{bmatrix}
0 & 0.5 & 0 & 0 & 0 & 0 & 0 \\
0.5 & 0 & 0.408 & 0 & 0 & 0 & 0.378 \\
0 & 0.408 & 0 & 0.447 & 0 & 0 & 0.378 \\
0 & 0 & 0.447 & 0 & 0.447 & 0 & 0 \\
0 & 0 & 0 & 0.447 & 0 & 0.447 & 0.378 \\
0 & 0 & 0 & 0 & 0.447 & 0 & 0.408 \\
0 & 0.378 & 0.378 & 0 & 0.378 & 0.408 & 0
\end{bmatrix}
$$

## 2.17  EXTENDED ADJACENCY MATRICES

The *extended vertex-adjacency matrix*, denoted by $\mathbf{E}^v\mathbf{A}$, is a square symmetric $V \times V$ matrix defined as (Yang et al., 1994)

$$
\left[\mathbf{E}^v\mathbf{A}\right]_{ij} = \begin{cases} \dfrac{[d(i)/d(j)]+[d(j)/d(i)]}{2} & \text{if vertices } i \text{ and } j \text{ are adjacent} \\ 0 & \text{otherwise} \end{cases} \tag{2.27}
$$

where $d(i)$ is the degree of a vertex $i$. This definition indicates that the $\mathbf{E}^v\mathbf{A}$ matrix is a sort of the edge-weighted vertex-adjacency matrix.

The extended vertex-adjacency matrix of $G_1$ (see structure $A$ in Figure 2.1) is presented as follows. The vertex-degrees in $G_1$ are given in Figure 2.17.

$$
\mathbf{E}^v\mathbf{A}(G_1) = \begin{bmatrix}
0 & 1.25 & 0 & 0 & 0 & 0 & 0 \\
1.25 & 0 & 1.08 & 0 & 0 & 0 & 0 \\
0 & 1.08 & 0 & 1.08 & 0 & 1.00 & 0 \\
0 & 0 & 1.08 & 0 & 1.00 & 0 & 0 \\
0 & 0 & 0 & 1.00 & 0 & 1.08 & 0 \\
0 & 0 & 1.00 & 0 & 1.08 & 0 & 1.67 \\
0 & 0 & 0 & 0 & 0 & 1.67 & 0
\end{bmatrix}
$$

The use of topological indices based on this matrix in QSPR is explored by Yang et al. (1994). However, these authors did not consider the *extended edge-adjacency matrix*, denoted by $\mathbf{E}^e\mathbf{A}$. The $\mathbf{E}^e\mathbf{A}$ matrix is based on the edge-degrees. Since the edge-degrees of a graph $G$ are equal to vertex-degrees of a line graph $L(G)$, it follows that

$$
\mathbf{E}^e\mathbf{A}(G) = \mathbf{E}^v\mathbf{A}\,(L(G)) \tag{2.28}
$$

From Equation (2.28) is also evident that the extended edge-adjacency matrix is also a sort of the edge-weighted adjacency matrix.

The extended edge-adjacency matrix of $G_1$ is equal to the extended vertex-adjacency matrix of $L(G_1)$.

$$\mathbf{E}^e\mathbf{A}(G_1) = \mathbf{E}^v\mathbf{A}[L(G_1)] = \begin{bmatrix} 0 & 1.25 & 0 & 0 & 0 & 0 & 0 \\ 1.25 & 0 & 1.00 & 0 & 0 & 0 & 1.04 \\ 0 & 1.00 & 0 & 1.08 & 0 & 0 & 1.04 \\ 0 & 0 & 1.08 & 0 & 1.08 & 0 & 0 \\ 0 & 0 & 0 & 1.08 & 0 & 1.08 & 1.04 \\ 0 & 0 & 0 & 0 & 1.08 & 0 & 1.25 \\ 0 & 1.04 & 1.04 & 0 & 1.04 & 1.25 & 0 \end{bmatrix}$$

Topological indices based on the extended edge-adjacency matrix have not yet been explored in QSPR or QSAR modeling.

## 2.18   ZAGREB MATRICES

The Zagreb matrices can also be considered the vertex- and edge-weighted matrices related to the vertex- and edge-connectivity matrices discussed above. They can be formulated in terms of the vertex- or edge-degrees.

### 2.18.1   ZAGREB MATRICES IN TERMS OF THE VERTEX-DEGREES

The *vertex-Zagreb matrix*, denoted by $^v\mathbf{ZM}$, is a diagonal $V \times V$ matrix defined in terms of the vertex-degrees $d$:

$$\left[ {}^v\mathbf{ZM} \right]_{ij} = \begin{cases} [d(i)]^2 & \text{if } i = j \\ 0 & \text{otherwise} \end{cases} \tag{2.29}$$

The $^v\mathbf{ZM}$ matrix for $G_1$ (see structure $A$ in Figure 2.1) is given below:

$$^v\mathbf{ZM}(G_1) = \begin{bmatrix} 1 & 0 & 0 & 0 & 0 & 0 & 0 \\ 0 & 4 & 0 & 0 & 0 & 0 & 0 \\ 0 & 0 & 9 & 0 & 0 & 0 & 0 \\ 0 & 0 & 0 & 4 & 0 & 0 & 0 \\ 0 & 0 & 0 & 0 & 4 & 0 & 0 \\ 0 & 0 & 0 & 0 & 0 & 9 & 0 \\ 0 & 0 & 0 & 0 & 0 & 0 & 1 \end{bmatrix}$$

Summation of the diagonal elements gives the first Zagreb index. This index is discussed in a number of reports (Gutman and Trinajstić, 1972, Gutman et al., 1975,

2005; Todeschini and Consonni, 2000, 2009; Nikolić et al., 2003a; Gutman and Das, 2004; Miličević et al., 2004; Zhou, 2004; Zhou and Gutman, 2005; Hansen and Vukičević, 2007; Janežič et al., 2007; Das and Trinajstić, 2011; Lučić et al., 2011).

The *modified* vertex-Zagreb matrix, denoted by $^{mv}\mathbf{ZM}$, is defined as (Miličević et al., 2004)

$$\left[ ^{mv}\mathbf{ZM} \right]_{ij} = \begin{cases} 1/[d(i)]^2 & \text{if } i = j \\ 0 & \text{otherwise} \end{cases} \tag{2.30}$$

The $^{mv}\mathbf{ZM}$ matrix for $G_1$ (see structure $A$ in Figure 2.1) is as follows:

$$^{mv}\mathbf{ZM}(G_1) = \begin{bmatrix} 1 & 0 & 0 & 0 & 0 & 0 & 0 \\ 0 & 1/4 & 0 & 0 & 0 & 0 & 0 \\ 0 & 0 & 1/9 & 0 & 0 & 0 & 0 \\ 0 & 0 & 0 & 1/4 & 0 & 0 & 0 \\ 0 & 0 & 0 & 0 & 1/4 & 0 & 0 \\ 0 & 0 & 0 & 0 & 0 & 1/9 & 0 \\ 0 & 0 & 0 & 0 & 0 & 0 & 1 \end{bmatrix}$$

Summation of the diagonal elements gives the *modified* first Zagreb index. It is also discussed in a number of reports (Gutman and Trinajstić, 1972, Gutman et al., 1975, 2005; Todeschini and Consonni, 2000, 2009; Nikolić et al., 2003a; Gutman and Das, 2004; Miličević et al., 2004; Zhou, 2004; Zhou and Gutman, 2005; Das and Trinajstić, 2011).

The *edge-Zagreb matrix*, denoted $^{e}\mathbf{ZM}$, is defined by

$$\left[ ^{e}\mathbf{ZM} \right]_{ij} = \begin{cases} d(i)d(j) & \text{if vertices } i \text{ and } j \text{ are adjacent} \\ 0 & \text{otherwise} \end{cases} \tag{2.31}$$

The edge-weighted Zagreb matrix $^{e}\mathbf{ZM}$ for $G_1$ (see structure $A$ in Figure 2.1) is presented below:

$$^{e}\mathbf{ZM}(G_1) = \begin{bmatrix} 0 & 2 & 0 & 0 & 0 & 0 & 0 \\ 2 & 0 & 6 & 0 & 0 & 0 & 0 \\ 0 & 6 & 0 & 6 & 0 & 9 & 0 \\ 0 & 0 & 6 & 0 & 4 & 0 & 0 \\ 0 & 0 & 0 & 4 & 0 & 6 & 0 \\ 0 & 0 & 9 & 0 & 6 & 0 & 3 \\ 0 & 0 & 0 & 0 & 0 & 3 & 0 \end{bmatrix}$$

Summation of the off-diagonal elements in the upper (or lower) matrix-triangle produces the second Zagreb index. This molecular descriptor is also discussed in a number of reports (Gutman and Trinajstić, 1972, Gutman et al., 1975, 2005; Todeschini and Consonni, 2000, 2009; Nikolić et al., 2003a; Gutman and Das, 2004; Das and Gutman, 2004; Miličević et al., 2004; Zhou, 2004; Zhou and Gutman, 2005; Das and Trinajstić, 2011).

The *modified* edge-Zagreb matrix, denoted by $^{me}\mathbf{ZM}$, is defined as

$$\left[ {}^{me}\mathbf{ZM} \right]_{ij} = \begin{cases} 1/[d(i)d(j)] & \text{if vertices } i \text{ and } j \text{ are adjacent} \\ 0 & \text{otherwise} \end{cases} \tag{2.32}$$

As an example, the modified edge-Zagreb matrix $^{me}\mathbf{ZM}$ for $G_1$ (see structure $A$ in Figure 2.1) is given below:

$$
{}^{me}\mathbf{ZM}(G_1) = \begin{bmatrix}
0 & 1/2 & 0 & 0 & 0 & 0 & 0 \\
1/2 & 0 & 1/6 & 0 & 0 & 0 & 0 \\
0 & 1/6 & 0 & 1/6 & 0 & 1/9 & 0 \\
0 & 0 & 1/6 & 0 & 1/4 & 0 & 0 \\
0 & 0 & 0 & 1/4 & 0 & 1/6 & 0 \\
0 & 0 & 1/9 & 0 & 1/6 & 0 & 1/3 \\
0 & 0 & 0 & 0 & 0 & 1/3 & 0
\end{bmatrix}
$$

Summation of the off-diagonal elements in the upper (or lower) matrix-triangle produces the *modified* second Zagreb index (Todeschini and Consonni, 2000, 2009; Nikolić et al., 2003a; Vukičević and Trinajstić, 2003; Miličević et al., 2004).

Finally, the *path-Zagreb matrix*, denoted by $^p\mathbf{ZM}$, has been introduced by Vukičević and his coworkers (Vukičević et al., 2009). The path-Zagreb matrix for trees $^p\mathbf{ZM}$ is defined as follows:

$$\left[ {}^{p}\mathbf{ZM} \right]_{ij} = \begin{cases} \displaystyle\prod_{i=1}^{N} d(i) & \text{if there is at least one path from } i \text{ to } j \\ 0 & \text{if there is no path from } i \text{ to } j \end{cases} \tag{2.33}$$

Below we give the path-Zagreb matrix for a branched tree $T_2$ representing the carbon skeleton of 2,3-dimethylhexane. In Figure 2.19, we give the labeled tree $T_2$ and the corresponding vertex-degrees.

$$
{}^{p}\mathbf{ZM}(T_2) = \begin{bmatrix}
0 & 3 & 9 & 18 & 36 & 36 & 9 & 3 \\
3 & 0 & 9 & 18 & 36 & 36 & 9 & 3 \\
9 & 9 & 0 & 6 & 12 & 12 & 3 & 9 \\
18 & 18 & 6 & 0 & 4 & 4 & 6 & 12 \\
36 & 36 & 12 & 4 & 0 & 2 & 12 & 36 \\
36 & 36 & 12 & 4 & 2 & 0 & 12 & 36 \\
9 & 9 & 3 & 6 & 12 & 12 & 0 & 9 \\
3 & 3 & 9 & 12 & 36 & 36 & 9 & 0
\end{bmatrix}
$$

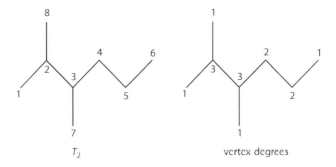

**FIGURE 2.19** Vertex-labels and vertex-degrees of branched tree $T_2$ representing the carbon skeleton of 2,3-dimethylhexane.

In graphs containing cycles, two vertices can be connected by more than one path. In this case we select the path with the smallest sum of vertex-weights. Below we give the path-Zagreb matrix for the graph $G_1$, whose vertex-degrees are given in Figure 2.17.

$$
{}^{p}\mathbf{ZM}(G_1) = \begin{bmatrix}
0 & 2 & 6 & 12 & 24 & 18 & 18 \\
2 & 0 & 6 & 12 & 24 & 18 & 18 \\
6 & 6 & 0 & 6 & 12 & 9 & 9 \\
12 & 12 & 6 & 0 & 4 & 12 & 12 \\
24 & 24 & 12 & 4 & 0 & 6 & 6 \\
18 & 18 & 9 & 12 & 6 & 0 & 3 \\
18 & 18 & 9 & 12 & 6 & 3 & 0
\end{bmatrix}
$$

### 2.18.2 ZAGREB MATRICES IN TERMS OF THE EDGE-DEGREES

It should be noted that the Zagreb matrices of a graph $G$ in terms of the edge-degrees are the vertex-Zagreb matrices of the corresponding line graph $L(G)$. The edge-degrees in $G_1$ and the vertex-degrees in $L(G_1)$ are given in Figure 2.18. The reader can easily confirm the above.

Zagreb indices found moderate use in structure-property modeling (Todeschini and Consonni, 2000, 2009; Das and Gutman, 2004; Miličević et al., 2004; Nikolić et al., 2004; Lučić et al., 2011). In this respect, a contribution by Peng et al. (2004) is important because they have shown how to improve the use of these indices in modeling molecular properties. Zagreb indices have also been used to study molecular complexity (Bertz and Wright, 1998; Nikolić et al., 2000a; 2000b; 2003), molecular chirality (Golbraikh et al., 2001), and isomerism (Golbraikh et al., 2002).

### 2.18.3 VARIABLE ZAGREB MATRICES IN TERMS OF THE VERTEX-DEGREES AND EDGE-DEGREES

Variable Zagreb matrices in terms of the vertex-degrees ${}^{v}\mathbf{VZM}$ and the edge-degrees ${}^{e}\mathbf{VZM}$ are given by (Miličević and Nikolić, 2004; Miličević et al., 2004; Nikolić et al., 2004):

$$\left[ {}^{v}\mathbf{VZM} \right]_{ij} = \begin{cases} [d(i)]^{2\lambda} & \text{if } i = j \\ 0 & \text{otherwise} \end{cases} \tag{2.34}$$

and

$$\left[ {}^{e}\mathbf{VZM} \right]_{ij} = \begin{cases} [d(i)d(j)]^{\lambda} & \text{if vertices } i \text{ and } j \text{ are adjacent} \\ 0 & \text{otherwise} \end{cases} \tag{2.35}$$

where $\lambda$ is the variable parameter. For $\lambda = 1$, (2.34) and (2.35) reduce to (2.29) and (2.31), respectively, and for $\lambda = -1$, to (2.30) and (2.32).

The variable Zagreb matrices are used to generate the variable Zagreb indices, which found application in structure-boiling point modeling and structure-molar volume modeling for aliphatic organic molecules (Miličević and Nikolić, 2004; Nikolić et al., 2004).

Recently, Gutman (2013) nicely reviewed the whole family of degree-based molecular descriptors, including the family of Zagreb indices. These graph-theoretical descriptors were tested against the standard heats of formations and normal boiling points of isomeric octanes producing fair descriptor-property correlations.

### 2.18.4  Zagreb Sum-Matrices

Another way to modify the edge-Zagreb matrix ${}^{e}\mathbf{ZM}$ is to sum up the degrees of vertices $i$ and $j$, making up bond $i$-$j$. The novel Zagreb matrix we call the *sum-edge-Zagreb matrix* and denote it by ${}^{se}\mathbf{ZM}$. It is defined as

$$\left[ {}^{se}\mathbf{ZM} \right]_{ij} = \begin{cases} d(i) + d(j) & \text{if vertices } i \text{ and } j \text{ are adjacent} \\ 0 & \text{otherwise} \end{cases} \tag{2.36}$$

An explicit example of this matrix is given below for the tree $T_2$ (Figure 2.7):

$${}^{se}\mathbf{ZM}(T_2) = \begin{bmatrix} 0 & 4 & 0 & 0 & 0 & 0 & 0 & 0 \\ 4 & 0 & 6 & 0 & 0 & 0 & 0 & 4 \\ 0 & 6 & 0 & 5 & 0 & 0 & 4 & 0 \\ 0 & 0 & 5 & 0 & 4 & 0 & 0 & 0 \\ 0 & 0 & 0 & 4 & 0 & 3 & 0 & 0 \\ 0 & 0 & 0 & 0 & 3 & 0 & 0 & 0 \\ 0 & 0 & 4 & 0 & 0 & 0 & 0 & 0 \\ 0 & 4 & 0 & 0 & 0 & 0 & 0 & 0 \end{bmatrix}$$

We give also the modified version of the sum-edge-Zagreb matrix, denoted by $^{mse}\mathbf{ZM}$. It is defined as follows:

$$
\left[ ^{mse}\mathbf{ZM} \right]_{ij} = \begin{cases} 1/[d(i)+d(j)] & \text{if vertices } i \text{ and } j \text{ are adjacent} \\ 0 & \text{otherwise} \end{cases} \tag{2.37}
$$

As an example of this matrix, we give below the $^{mse}\mathbf{ZM}$ matrix of the tree $T_2$:

$$
^{mse}\mathbf{ZM}(T_2) = \begin{bmatrix}
0 & 1/4 & 0 & 0 & 0 & 0 & 0 & 0 \\
1/4 & 0 & 1/6 & 0 & 0 & 0 & 0 & 1/4 \\
0 & 1/6 & 0 & 1/5 & 0 & 0 & 1/4 & 0 \\
0 & 0 & 1/5 & 0 & 1/4 & 0 & 0 & 0 \\
0 & 0 & 0 & 1/4 & 0 & 1/3 & 0 & 0 \\
0 & 0 & 0 & 0 & 1/3 & 0 & 0 & 0 \\
0 & 0 & 1/4 & 0 & 0 & 0 & 0 & 0 \\
0 & 1/4 & 0 & 0 & 0 & 0 & 0 & 0
\end{bmatrix}
$$

Finally, the sum-edge-Zagreb matrix $^{se}\mathbf{ZM}$ and modified sum-edge-Zagreb matrix $^{mse}\mathbf{ZM}$ can be collected as the variable sum-Zagreb matrices $^{se}\mathbf{VZM}$:

$$
\left[ ^{se}\mathbf{VZM} \right]_{ij} = \begin{cases} [d(i)+d(j)]^{\lambda} & \text{if vertices } i \text{ and } j \text{ are adjacent} \\ 0 & \text{otherwise} \end{cases} \tag{2.38}
$$

where $\lambda$ is the variable parameter. For $\lambda = 1$, (2.38) reduces to (2.36), and for $\lambda = -1$, reduces to (2.37).

## 2.19   THE HÜCKEL MATRIX

The *Hückel matrix* encountered in the Hückel theory of conjugated systems (Hückel, 1931, 1932a, 1932b) may be considered to be an augmented vertex-adjacency matrix (Trinajstić, 1992). In the Hückel theory, the aim is to solve the following secular equation set up for a given conjugated system (Streitwieser, 1961):

$$
\det |-e_i \mathbf{S} + \mathbf{H}| = 0; \; i = 1, \dots, V \tag{2.39}
$$

where $\mathbf{H}$ is the Hamiltonian matrix, $\mathbf{S}$ is the overlap matrix, $e_i$ is the set of eigenvalues, and $V$ is the number of $\pi$-electrons in a conjugated molecule. The Hamiltonian matrix and the overlap matrix in the Hückel theory are simplified by using the set of approximations originally introduced by Bloch (1929, 1930), but known in the quantum-chemical literature as the Hückel approximations (Streitwieser, 1961):

$$\mathbf{H} = \alpha \, \mathbf{I} + \beta \, {}^v\mathbf{A} \qquad (2.40)$$

$$\mathbf{S} = \mathbf{I} \qquad (2.41)$$

where $\mathbf{I}$ is the unit $V \times V$ matrix, $\alpha$ is an atomic parameter (atomic Coulomb integral), $\beta$ is a bond parameter (resonance integral), and ${}^v\mathbf{A}$ is the vertex-adjacency matrix.

Introducing (2.40) and (2.41) into (2.39), we obtain:

$$\det |(-e_i + \alpha) \, \mathbf{I} + \beta \, {}^v\mathbf{A}| = 0; \; i = 1, \, ..., \, V \qquad (2.42)$$

where the $e_i$ are the Hückel eigenvalues. If $\beta$ is used as the unit of energy and $\alpha$ as the zero-reference point, then the determinant (2.42) reduces to

$$\det |(-e_i) \, \mathbf{I} + {}^v\mathbf{A}| = 0; \; i = 1, \, ..., \, V \qquad (2.43)$$

Hence, the Hückel matrix $\mathbf{X}$ is a kind of augmented vertex-adjacency matrix:

$$\mathbf{X} = -e_i \, \mathbf{I} + {}^v\mathbf{A}; \; i = 1, \, ..., \, V \qquad (2.44)$$

In Figure 2.20, we give graph $G_6$ representing the carbon skeleton of 1,2-divinylcyclobutadiene.

The corresponding Hückel matrix is given by

$$\mathbf{X}(G_6) = \begin{bmatrix} -e & 1 & 0 & 0 & 0 & 0 & 0 & 0 \\ 1 & -e & 1 & 0 & 0 & 0 & 0 & 0 \\ 0 & 1 & -e & 1 & 0 & 1 & 0 & 0 \\ 0 & 0 & 1 & -e & 1 & 0 & 0 & 0 \\ 0 & 0 & 0 & 1 & -e & 1 & 0 & 0 \\ 0 & 0 & 1 & 0 & 1 & -e & 1 & 0 \\ 0 & 0 & 0 & 0 & 0 & 1 & -e & 1 \\ 0 & 0 & 0 & 0 & 0 & 0 & 1 & -e \end{bmatrix}$$

The Hückel theory was widely used in the early days of quantum chemistry, before the age of computers, and in spite of its simplicity, it is still in use; e.g., papers appear continually in *Physics Reviews B* using the Hückel theory in a solid-state

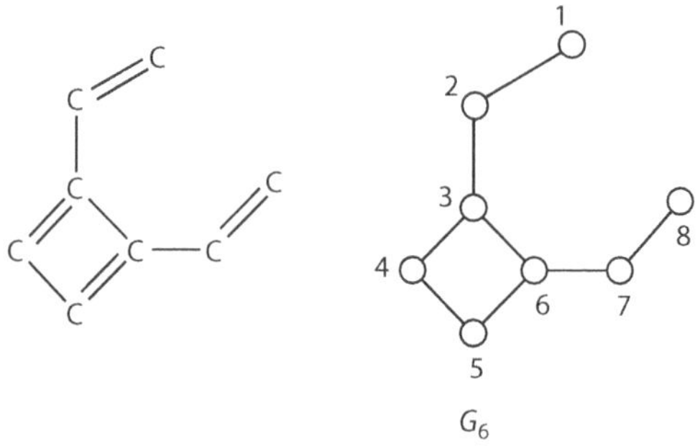

FIGURE 2.20   Graph $G_6$ representing the carbon skeleton of 1,2-divinylcyclobutadiene.

tight-binding format. The Hückel theory also plays a central role in a paper by Klein and Misra (2004) in their discussion of minimally Kekulenoid $\pi$-networks and reactivity of acyclics. Additionally, graph-theoretical analysis of the Hückel theory (e.g., Trinajstić, 1977a, 1977b; Graovac et al., 1977; Hosoya, 1999, 2003) has provided elegant answers to some questions that were previously left unanswered, such as why only a few (six to be precise (Cvetković et al., 1974b)) conjugated molecules have only integers as eigenvalues, or why some structurally quite different molecules possess identical sets of eigenvalues (that is, identical Hückel spectra) (Gutman and Trinajstić, 1973; Herndon, 1974; Živković et al., 1975; Herndon and Ellzey, 1975, 1986, Randić et al., 1976, 1989, D'Amato et al., 1981, Trinajstić, 1983), etc. Graphs with identical spectra are called *isospectral* or *cospectral* graphs. The first report on isospectral graphs was by Collatz and Sinogowitz (1957). For example, we give in Figure 2.21 graphs $G_7$ and $G_8$ of two structurally different molecules, 1,4-divinylbenzene and 2-phenylbutadiene, that possess identical Hückel spectra. Their vertex-adjacency matrices are given below Figure 2.21.

$$
{}^{v}\mathbf{A}(G_7) = \begin{bmatrix}
0 & 1 & 0 & 0 & 0 & 0 & 0 & 0 & 0 & 0 \\
1 & 0 & 1 & 0 & 0 & 0 & 0 & 0 & 0 & 0 \\
0 & 1 & 0 & 1 & 0 & 0 & 0 & 1 & 0 & 0 \\
0 & 0 & 1 & 0 & 1 & 0 & 0 & 0 & 0 & 0 \\
0 & 0 & 0 & 1 & 0 & 1 & 0 & 0 & 0 & 0 \\
0 & 0 & 0 & 0 & 1 & 0 & 1 & 0 & 1 & 0 \\
0 & 0 & 0 & 0 & 0 & 1 & 0 & 1 & 0 & 0 \\
0 & 0 & 1 & 0 & 0 & 0 & 1 & 0 & 0 & 0 \\
0 & 0 & 0 & 0 & 0 & 1 & 0 & 0 & 0 & 1 \\
0 & 0 & 0 & 0 & 0 & 0 & 0 & 0 & 1 & 0
\end{bmatrix}
$$

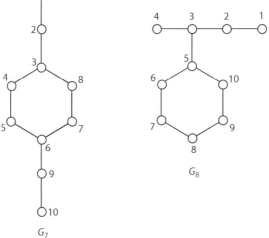

$G_7$

$G_8$

**FIGURE 2.21** A classical example of a pair of isospectral graphs.

$$^{v}\mathbf{A}(G_8) = \begin{bmatrix} 0 & 1 & 0 & 0 & 0 & 0 & 0 & 0 & 0 & 0 \\ 1 & 0 & 1 & 0 & 0 & 0 & 0 & 0 & 0 & 0 \\ 0 & 1 & 0 & 1 & 1 & 0 & 0 & 0 & 0 & 0 \\ 0 & 0 & 1 & 0 & 0 & 0 & 0 & 0 & 0 & 0 \\ 0 & 0 & 1 & 0 & 0 & 1 & 0 & 0 & 0 & 1 \\ 0 & 0 & 0 & 0 & 1 & 0 & 1 & 0 & 0 & 0 \\ 0 & 0 & 0 & 0 & 0 & 1 & 0 & 1 & 0 & 0 \\ 0 & 0 & 0 & 0 & 0 & 0 & 1 & 0 & 1 & 0 \\ 0 & 0 & 0 & 0 & 0 & 0 & 0 & 1 & 0 & 1 \\ 0 & 0 & 0 & 0 & 1 & 0 & 0 & 0 & 1 & 0 \end{bmatrix}$$

The corresponding Hückel spectra are identical: $\{G_7\} = \{G_8\} = \{\pm 2.2143, \pm 1.6751, \pm 1.0000, \pm 1.0000, \pm 0.5392\}$.

## 2.20   THE LAPLACIAN MATRIX

The *Laplacian matrix*, denoted by $\mathbf{L}$, is a real symmetric $V \times V$ matrix that may also be considered a kind of augmented vertex-adjacency matrix. It is defined as the following *difference matrix* (Mohar, 1989):

$$\mathbf{L} = \mathbf{\Delta} - {}^{v}\mathbf{A} \tag{2.45}$$

where $\mathbf{\Delta}$ is a *diagonal matrix* of dimension $V \times V$ whose diagonal entries are the vertex-degrees:

$$[\mathbf{\Delta}]_{ij} = \begin{cases} d(i) & \text{if } i = j \\ 0 & \text{otherwise} \end{cases} \tag{2.46}$$

where $d(i)$ is the degree of a vertex $i$. This matrix is also called the *vertex-degree matrix* (Todeschini and Consonni, 2000, 2009).

The entries of the Laplacian matrix are as follows:

$$[\mathbf{L}]_{ij} = \begin{cases} -1 & \text{if vertices } i \text{ and } j \text{ are adjacent} \\ d(i) & \text{if } i = j \\ 0 & \text{otherwise} \end{cases} \tag{2.47}$$

It should be noted that the smallest eigenvalue of $\mathbf{L}$ is always equal to zero, as a consequence of the special structure of the Laplacian matrix.

The Laplacian matrix is sometimes also called the *Kirchhoff matrix* (Mohar, 1989; Kunz, 1992, 1993, 1995) due to its role in the *matrix-tree theorem* (Cvetković et al., 1995), implicit (Moon, 1970) in the electrical network work of Kirchhoff (1847; in his paper Kirchhoff also introduced the concept of the *spanning tree*, though he did not use this term). It is also known as the *admittance matrix* (Cvetković et al., 1995). However, the name *Laplacian matrix* appears to be more appropriate since this matrix is just the matrix of a discrete Laplacian operator, which is one of the basic differential operators in quantum chemistry and beyond. Below we give the Laplacian matrix of the vertex-labeled graph $G_1$ (see structure $A$ in Figure 2.1):

$$\mathbf{L}(G_1) = \begin{bmatrix} 1 & -1 & 0 & 0 & 0 & 0 & 0 \\ -1 & 2 & -1 & 0 & 0 & 0 & 0 \\ 0 & -1 & 3 & -1 & 0 & -1 & 0 \\ 0 & 0 & -1 & 2 & -1 & 0 & 0 \\ 0 & 0 & 0 & -1 & 2 & -1 & 0 \\ 0 & 0 & -1 & 0 & -1 & 3 & -1 \\ 0 & 0 & 0 & 0 & 0 & -1 & 1 \end{bmatrix}$$

The Laplacian matrix is used to enumerate the number of spanning trees (e.g., Trinajstić et al., 1994; Nikolić et al., 1996). Let us remind the reader that a *spanning tree* of a graph $G$ is a connected acyclic subgraph containing all the vertices of $G$ (Harary, 1971).

If a graph contains a single cycle, then the number of spanning trees is simply equal to the size of the cycle. Thus, since the graph $G_1$ contains a four-membered cycle (see Figure 1.1), it possesses four spanning trees, denoted by $ST_n$ ($n = 1, 2, 3, 4$). They are depicted in Figure 2.22, and will be utilized later on in Section 4.17.

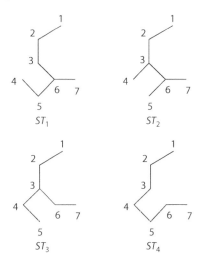

**FIGURE 2.22** Labeled spanning trees of $G_1$.

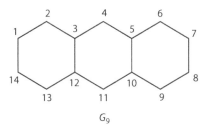

**FIGURE 2.23**  Labeled anthracene graph.

When a polycyclic graph $G$ is considered, obtaining the number of spanning trees is rather involved. One needs first to compute the Laplacian spectrum, and then to use it in the following counting formula, based on the matrix-tree theorem, in order to get the number of spanning trees of $G$ (Mohar, 1989):

$$t(G) = (1/V) \prod_{i=2}^{V} \lambda_i \tag{2.48}$$

where $t(G)$ is the number of spanning trees of $G$ and $\lambda_i$ ($i = 2, ..., V$) are the eigenvalues of the Laplacian matrix. There is a reason why the multiplication of eigenvalues in this formula starts with $\lambda_2$: the smallest member of the Laplacian spectrum $\lambda_1$ is always zero.

We will exemplify the use of formula (2.48) to enumerate the spanning trees of a graph representing the carbon skeleton of anthracene. This graph is shown in Figure 2.23.

The Laplacian spectrum of the anthracene graph is {0.0000, 0.1981, 0.7530, 0.8402, 1.1206, 1.5500, 1.6207, 2.4450, 3.2470, 3.3473, 3.4919, 3.8019, 4.5321, 5.0472}. Now, we enter these numbers in formula (2.48) to get the number of spanning trees of the anthracene graph: 204. Spanning trees have found use in ring current calculations on conjugated systems (Mallion, 1975) and in assessing the complexity of molecules and graphs, and reaction mechanisms (e.g., Trinajstić, 1994; Temkin et al., 1996; Nikolić et al., 1996b, 1999, 2000, 2003; Nikolić and Trinajstić, 2003). In the next section, we will show a simpler approach to the enumeration of the number of spanning trees of polycyclic graphs.

Several topological indices based on the Laplacian matrix have been proposed, besides the number of spanning trees (Mohar, 1989; Kirby et al., 2004), the Mohar indices (Mohar, 1989; Trinajstić et al., 1994), the Wiener index of trees (Mohar, 1989, 1991; Mohar et al., 1993; Trinajstić et al., 1994), the *quasi*-Wiener index (Marković et al., 1995), and spanning tree density and reciprocal spanning tree density (Mallion and Trinajstić, 2003).

## 2.21  THE GENERALIZED LAPLACIAN MATRIX

Gutman et al. (1983) proposed a simple method for computing the number of spanning trees of planar polycyclic graphs. It is based on the vertex-weighted inner dual

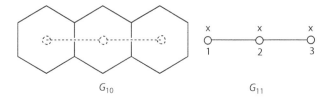

**FIGURE 2.24** Anthracene graph $G_{10}$ and its vertex-weighted inner dual $G_{11}$.

of polycyclic graphs. An *inner dual G'* of a planar polycyclic graph $G$ is obtained by placing a vertex in each cycle of $G$, and a pair of vertices in $G'$ is connected if the corresponding cycles in $G$ have an edge in common. In a vertex-weighted inner dual, vertices are weighted by the numbers representing the sizes of the corresponding cycles. In Figure 2.24, we give the anthracene graph $G_{10}$ and its weighted inner dual $G_{11}$ with weights denoted by $x$.

The generalized Laplacian matrix, that is, the Laplacian matrix of the weighted inner dual, is given by

$$\mathbf{L} = [{}^w\mathbf{\Delta} - {}^v\mathbf{A}] \qquad (2.49)$$

where ${}^w\mathbf{\Delta}$ is the weighted diagonal matrix and ${}^v\mathbf{A}$ is the vertex-adjacency matrix of the vertex-weighted inner dual. The determinant of the generalized Laplacian matrix gives the polynomial of the weighted inner dual. Substituting the values of vertex-weights by the cycle sizes, one obtains the number of spanning trees.

The application of this procedure is demonstrated below for the anthracene graph and its weighted inner dual. The ${}^w\mathbf{\Delta}$ and ${}^v\mathbf{A}$ matrices of the weighted inner dual $G_{11}$ are as follows:

$$
{}^w\mathbf{\Delta}(G_{11}) = \begin{bmatrix} x & 0 & 0 \\ 0 & x & 0 \\ 0 & 0 & x \end{bmatrix}
$$

$$
{}^v\mathbf{A}(G_{11}) = \begin{bmatrix} 0 & 1 & 0 \\ 1 & 0 & 1 \\ 0 & 1 & 0 \end{bmatrix}
$$

and the corresponding generalized Laplacian polynomial of $G_{11}$ is

$$
\det|\mathbf{L}(G_{11})| = \det \begin{vmatrix} x & -1 & 0 \\ -1 & x & -1 \\ 0 & -1 & x \end{vmatrix} = x^3 - 2x \qquad (2.50)
$$

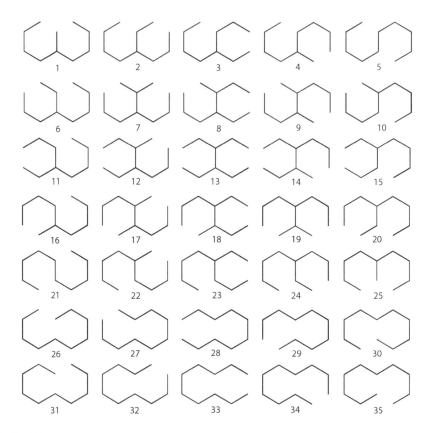

**FIGURE 2.25**  All spanning trees for the graph representing the carbon skeleton of naphthalene.

Substituting 6 for x in (2.50), the number of spanning trees obtained for the anthracene graph is 204. In the case of naphthalene, both rings are six-membered, and the number of spanning trees is much smaller, that is, $x^2 - 1 = 35$. Spanning trees of naphthalene are depicted in Figure 2.25.

To demonstrate how the cycle sizes influence the number of spanning trees of polycyclic graphs, we compute this number for a tricyclic graph corresponding to benzo[$f$]azulene. This graph and its vertex-weighted inner dual are shown in Figure 2.26.

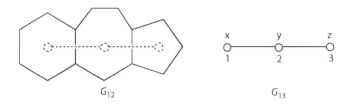

**FIGURE 2.26**  Benzo[$f$]azulene graph $G_{12}$ and its vertex-weighted inner dual $G_{13}$.

The generalized Laplacian polynomial of the inner dual of the benzo[*f*]azulene graph is given by

$$\det|\mathbf{L}(G_{13})| = \det\begin{vmatrix} x & -1 & 0 \\ -1 & y & -1 \\ 0 & -1 & z \end{vmatrix} = xyz - x - z \tag{2.51}$$

The sizes of cycles in the benzo[*f*]azulene graph are $x = 6$, $y = 7$, and $z = 5$. Substitution of these numbers in the above polynomial gives the number of spanning trees of $G_{12}$ as 199. In the case of azulene, the sizes of cycles are $y = 7$ and $z = 5$, and the number of spanning trees is $zy - 1 = 34$. They are depicted in Figure 2.27.

This approach by Gutman et al. (1983) for counting spanning trees applies only to planar polycyclic graphs. Later, Kirby et al. (2004) put forward a theorem for counting spanning trees in general molecular graphs, that is, nonplanar graphs with loops

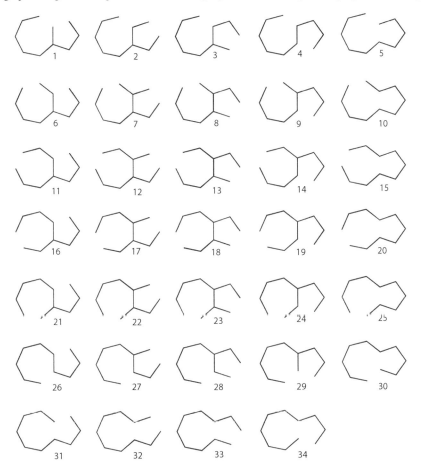

**FIGURE 2.27** All spanning trees for the graph representing the carbon skeleton of azulene.

and multiple edges, with particular application to toroidal fullerenes. The approach by Kirby et al. (2004) is based on the following counting formula:

$$t(G) = \det \mathbf{V}/(\det \mathbf{U})^2 \qquad (2.52)$$

where the $\mathbf{V}$ matrix is defined as

$$\mathbf{V} = (\mathbf{CE})(\mathbf{CE})^{\mathrm{T}} \qquad (2.53)$$

and the $\mathbf{U}$ matrix is a nonsingular matrix chosen in such a way that its determinant is 1 or some small integer. The $\mathbf{CE}$ matrix is the cycle-edge incidence matrix, the inverse of the edge-cycle incidence matrix (see Section 3.4).

Kirby et al. (2004) considered three kinds of cycles—independent cycles, fundamental cycles, and patch cycles—and showed that they all produce the same number of spanning trees. However, det $\mathbf{U}$ is equal to 1 only for fundamental cycles of any graph and for patch cycles of any planar graph. It should be noted that the procedure set up by Kirby et al. (2004) reduces for planar graphs to that introduced by Gutman et al. (1983).

## 2.22   THE AUGMENTED VERTEX-DEGREE MATRIX

The *augmented vertex-degree matrix*, denoted $^a\mathbf{\Delta}$, is an unsymmetric $V \times V$ matrix defined as (Randić, 2001a; Randić and Plavšić, 2002, 2003)

$$\left[ ^a\mathbf{\Delta} \right]_{ij} = \begin{cases} d(j)/2^{l(i,j)} & \text{if } i \neq j \\ d(i) & \text{if } i = j \end{cases} \qquad (2.54)$$

where $d(j)$ is the degree of a vertex $j$ and $l(i,j)$ is the distance between vertices $i$ and $j$.

As an illustrative example of the augmented vertex-degree matrix, we give below this matrix for $G_1$ (see structure $A$ in Figure 2.1) whose vertex-degrees are given in Figure 2.17.

$$
^a\mathbf{\Delta}(G_1) =
\begin{bmatrix}
1 & 2/2 & 3/4 & 2/8 & 2/16 & 3/8 & 1/16 \\
1/2 & 2 & 3/2 & 2/4 & 2/8 & 3/4 & 1/8 \\
1/4 & 2/2 & 3 & 2/2 & 2/4 & 3/2 & 1/4 \\
1/8 & 2/4 & 3/2 & 2 & 2/2 & 3/4 & 1/8 \\
1/16 & 2/8 & 3/4 & 2/2 & 2 & 3/2 & 1/4 \\
1/8 & 2/4 & 3/2 & 2/4 & 2/2 & 3 & 2/2 \\
1/16 & 2/8 & 3/4 & 2/8 & 2/4 & 3/2 & 1
\end{bmatrix}
$$

The augmented vertex-degree matrix can be used to compute the complexity index proposed by Randić (2001a) and Randić and Plavšić (2002, 2003). The Randić-Plavšić complexity index equals the sum of all the matrix row-sums for vertices nonequivalent by symmetry. It should be noted that the $i$-th row-sum represents the augmented degree of the vertex $i$.

Randić (2001a) and Randić and Plavšić (2002, 2003) studied with their approach complexity of acyclic and cyclic graphs and transitive graphs representing degenerate rearrangements.

## 2.23 DISTANCE-WEIGHTED ADJACENCY MATRIX

Randić and his coworkers (Randić et al., 2010) introduced the *distance-weighted adjacency matrix*, denoted by **DWA**, as a part of their project on developing the *natural distance matrices*. The **DWA** matrix for acyclic graphs is defined as

$$[\mathbf{DWA}]_{ij} = \begin{cases} [d(i) + d(j)]^{1/2} & \text{if vertices } i \text{ and } j \text{ are adjacent} \\ 0 & \text{otherwise} \end{cases} \tag{2.55}$$

An example of this matrix is given below for the tree $T_2$ (see Figure 2.19):

$$\mathbf{DWA}(T_2) = \begin{bmatrix} 0 & 2 & 0 & 0 & 0 & 0 & 0 & 0 \\ 2 & 0 & 3 & 0 & 0 & 0 & 0 & 2 \\ 0 & 3 & 0 & \sqrt{5} & 0 & 0 & 2 & 0 \\ 0 & 9 & \sqrt{5} & 0 & 2 & 0 & 0 & 0 \\ 0 & 0 & 0 & 2 & 0 & \sqrt{3} & 0 & 0 \\ 0 & 0 & 0 & 0 & \sqrt{3} & 0 & 0 & 0 \\ 0 & 0 & 2 & 0 & 0 & 0 & 0 & 0 \\ 0 & 2 & 0 & 0 & 0 & 0 & 0 & 0 \end{bmatrix}$$

By substituting $[d(i) + d(j)]^{1/2}$ with $[d(i) + d(j)]^{-1/2}$ in (2.55), one obtains the sum-vertex-connectivity matrix (Zhou and Trinajstić, 2009, 2010a; Randić et al., 2010); see Section 2.15. If one substitutes $[d(i) + d(j)]^{1/2}$ with either $[d(i) + d(j)]$ or $[d(i) + d(j)]^{-1}$, the sum-edge-Zagreb matrix or modified-sum-edge-Zagreb matrix is obtained, respectively. Thus, it appears that all these matrices can be traced to the Randić connectivity matrix proposed years ago (Randić, 1992).

## REFERENCES

R. Aringhieri, P. Hansen, and F. Malucelli, Chemical trees enumeration algorithms, Università di Paoia, Dipartimento di Informatice, Pisa, 1999.

A.A. Balandin, Structural algebra in chemistry, *Acta Physicochim. USSR* 12 (1940) 447–479.

K. Balasubramanian, Applications of edge-colorings of graphs and characteristic polynomials to spectroscopy and quantum chemistry, *J. Mol. Struct.* (*Theochem*) 185 (1989) 229–248.

M. Barysz, D. Plavšić, and N. Trinajstić, A note on topological indices, *MATCH Commun. Math. Comput. Chem.* 19 (1986) 89–116.

S.C. Basak, S. Nikolić, N. Trinajstić, D. Amić, and D. Bešlo, QSPR modeling: Graph connectivity indices versus line graph connectivity indices, *J. Chem. Inf. Comput. Sci.* 40 (2000) 927–933.

S.H. Bertz and W.F. Wright, The graph theory approach to synthetic analysis: Definition and application of molecular complexity and synthetic complexity, *Graph Theory Notes NY* 35 (1998) 32–48.

F. Bloch, Über die Quantenmechanik der Elektronen in Kristallgittern, *Z. Phys.* 52 (1929) 555–600.

F. Bloch, Zur Theorie des Ferromagnetismus, *Z. Phys.* 61 (1930) 206–219.

D. Bonchev, Overall connectivity—a next generation molecular connectivity, *J. Mol. Graphics Modell.* 20 (2001) 65–75.

D. Bonchev, Overall connectivity and topological complexity: A new tool for QSPR/QSAR, in J. Devillers and A.T. Balaban, Eds., *Topological Indices and Related Descriptors*, Gordon and Breac, Reading, U.K., 1999, pp. 361–401.

D. Bonchev, On the complexity of Platonic solids, *Croat. Chem. Acta* 77 (2004) 167–173.

D. Bonchev and L.B. Kier, Topological atomic indices and the electronic changes in alkanes, *J. Math. Chem.* 9 (1992) 75–85.

D. Bonchev and N. Trinajstić, Overall molecular descriptors. 3. Overall Zagreb indices, *SAR QSAR Environ. Res.* 12 (2001) 213–236.

F.R. Burden, Molecular identification number for substructure searches, *J. Chem. Inf. Comput. Sci.* 29 (1989) 225–227.

S. Carter, N. Trinajstić, and S. Nikolić, A note on the use of ID numbers in QSAR studies, *Acta Pharm. Jugosl.* 37 (1987) 37–42.

G.G. Cash, A fast computer algorithm for finding the permanent of adjacency matrices, *J. Math. Chem.* 18 (1995) 115–119.

L. Collatz and U. Sinogowitz, Spektren endlicher Graphen, *Abh. Math. Sem. Univ. Hamburg* 21 (1957) 63–77.

V. Consonni and R. Todeschini, Multivariate analysis of molecular descriptors, in *Statistical modelling of molecular descriptors in QSAR/QSPR*, ed. M. Dehmer, K. Varmuza, and D. Bonchev, Wiley-Blackwell, Weinheim, 2012, pp. 111–147.

C.A. Coulson and G.S. Rushbroke, Note on the method of molecular orbitals, *Proc. Camb. Phil. Soc.* 36 (1940) 193–200.

D. Cvetković, M. Doob, and H. Sachs, *Spectra of graphs—Theory and applications*, 3rd ed., Johann Ambrosius Barth Verlag, Heidelberg, 1995.

D.M. Cvetković and I. Gutman, Note on branching, *Croat. Chem. Acta* 49 (1977) 115–121.

D. Cvetković, I. Gutman, and N. Trinajstić, Kekulé structures and topology, *Chem. Phys. Lett.* 16 (1972) 614–616.

D. Cvetković, I. Gutman, and N. Trinajstić, Graph theory and molecular orbitals. VII. The role of resonance structures, *J. Chem. Phys.* 61 (1974a) 2700–2706.

D.M. Cvetković, I. Gutman, and N. Trinajstić, Conjugated molecules having integral spectra, *Chem. Phys. Lett.* 29 (1974b) 65–68.

S.S. D'Amato, B.M. Gimarc, and N. Trinajstić, Isospectral and subspectral molecules, *Croat. Chem. Acta* 54 (1981) 1–52.

K.C. Das and I. Gutman, Some properties of the second Zagreb index, *MATCH Commun. Math. Comput. Chem.* 52 (2004) 103–112.

K.C. Das and N. Trinajstić, Relationship between the eccentric connectivity index and Zagreb indices, *Comput. Math. Appl.* 62 (2011) 1758–1764.

J. Devillers and A.T. Balaban, eds., *Topological indices and related descriptors in QSAR and QSPR*, Gordon & Breach, Amsterdam, 1999.

M.V. Diudea, Ed., QSPR/QSAR Studies by Molecular Descriptors, Nova, Huntington, NY, 2001.

Z. Du, B. Zhou, and N. Trinajstić, Minimum sum-connectivity indices of trees and unicyclic graphs of a given matching number, *J. Math. Chem.* 47 (2010) 842–855.

Z. Du, B. Zhou, and N. Trinajstić, On the general sum-connectivity index of trees, *Appl. Math. Lett.* 24 (2011) 402–405.

J. Dugundji and I. Ugi, An algebraic model for constitutional chemistry as a basis for chemical computer programs, *Topics Curr. Chem.* 39 (1973) 19–64.

E. Estrada, Edge adjacency relationships and a novel topological index related to molecular volume, *J. Chem. Inf. Comput. Sci.* 35 (1995a) 31–33.

E. Estrada, Edge adjacency relationships in molecular graphs containing heteroatoms: A new topological index related to molecular volume, *J. Chem. Inf. Comput. Sci.* 35 (1995b) 701–707.

E. Estrada, Graph theoretical invariant of Randić revisited, *J. Chem. Inf. Comput. Sci.* 35 (1995c) 1022–1025.

E. Estrada, Novel strategies in the search of topological indices, in *Topological indices and related descriptors in QSAR and QSPR*, ed. J. Devillers and A.T. Balaban, Gordon & Breach, Amsterdam, 1999, pp. 403–453.

R. Garcia-Domenech, J. Gálvez, J.V. De Julián-Ortiz, and L. Pogliani, Some new trends in chemical graph theory, *Chem. Rev.* 108 (2008) 1127–1169.

D.J. Gluck, A chemical structure storage and search system developed at Du Pont, *J. Chem. Doc.* 5 (1965) 43–51.

A. Golbraikh, D. Bonchev, and A. Tropsha, Novel chirality descriptors derived from molecular topology, *J. Chem. Inf. Comput. Sci.* 41 (2001) 147–158.

A. Golbraikh, D. Bonchev, and A. Tropsha, Novel ZE-isomerism descriptors derived from molecular topology and their application to QSAR analysis, *J. Chem. Inf. Comput. Sci.* 42 (2002) 769–787.

M. Gordon and G.R. Scantlebury, Non-random polycondensation statistical theory of the substitution effect, *Trans Faraday Soc.* 60 (1964) 604–621.

A. Graovac, I. Gutman, and N. Trinajstić, *Topological approach to the chemistry of conjugated molecule*, Springer, Berlin, 1977.

A. Graovac, O.E. Polansky, N. Trinajstić, and N. Tyutyulkov, Graph theory in chemistry. II. Graph-theoretical description of heteroconjugated molecules, *Z. Naturforsch.* 30a (1975) 1696–1699.

A. Graovac and N. Trinajstić, Graphs in chemistry, *MATCH Commun. Math. Comput. Chem.* 1 (1975a) 159–170.

A. Graovac and N. Trinajstić, Möbius molecules and graphs, *Croat. Chem. Acta* 47 (1975b) 95–104.

A. Graovac and N. Trinajstić, Graphical description of Möbius molecules, *J. Mol. Struct.* 30 (1976) 416–420.

I. Gutman, Electronic properties of Möbius systems, *Z. Naturforsch.* 33a (1978) 214–216.

I. Gutman, Molecular graphs with minimal and maximal Randić indices, *Croat. Chem. Acta* 75 (2002) 357–369.

I. Gutman, Degree-based topological indices, *Croat. Chem. Acta* 86 (2013) 351–361.

I. Gutman and K.C. Das, The first Zagreb index 30 years after, *MATCH Commun. Math. Comput. Chem.* 50 (2004) 83–92.

I. Gutman and E. Estrada, Topological indices based on the line graph of the molecular graph, *J. Chem. Inf. Comput. Sci.* 36 (1996) 541–543.

I. Gutman, B. Furtula, A.A. Toropov, and A.P. Toropova, The graph of atomic orbitals and its basic properties. 2. Zagreb indices, *MATCH Commun. Math. Comput. Chem.* 53 (2005) 225–230.

I. Gutman and B. Furtula, eds., *Recent results in the theory of Randić index, Mathematical Chemistry Monographs, MCM-6,* University of Kragujevac and Faculty of Science, Kragujevac, Serbia, 2008.

I. Gutman, R.B. Mallion, and J.W. Essam, Counting spanning trees of a labelled molecular-graph, *Mol. Phys.* 50 (1983) 859–877.

I. Gutman, L.J. Petrović, E. Estrada, and S.H. Bertz, The line graph model. Predicting physico-chemical properties of alkanes, *ACH-Models Chem.* 135 (1998) 147–155.

I. Gutman and O.E. Polansky, *Mathematical concepts in organic chemistry,* Springer, Berlin, 1986.

I. Gutman, G. Rücker, and C. Rücker, On walks in molecular graphs, *J. Chem. Inf. Comput. Sci.* 41 (2001) 739–745.

I. Gutman, B. Ruščić, N. Trinajstić, and C.F. Wilcox Jr., Graph theory and molecular orbitals. XII. Acyclic polyenes, *J. Chem. Phys.* 62 (1975) 3399–3405.

I. Gutman and N. Trinajstić, Graph theory and molecular orbitals. Total $\pi$-electron energy of alternant hydrocarbons, *Chem. Phys. Lett.* 17 (1972) 535–538.

I. Gutman and N. Trinajstić, Graph theory and molecular orbitals, *Topics Curr. Chem.* 42 (1973) 49–93.

P. Hansen and D. Vukičević. Comparing the Zagreb indices, *Croat. Chem. Acta* 80 (2007) 165–168.

F. Harary, *Graph Theory,* 2nd printing, Addison-Wesley, Reading, MA, 1971.

E. Heilbronner, Hückel molecular orbitals of Möbius-type conformations of annulenes, *Tetrahedon Lett.* (1964) 1923–1928.

R. Herges, Topology in chemistry: Designing Möbius molecules, *Chem. Rev.* 106 (2006) 4820–4842.

W.C. Herndon, Isospectral molecules, *Tetrahedron Lett.* (1974) 671–674.

W.C. Herndon and M.L. Ellzey Jr., Isospectral graphs and molecules, *Tetrahedron* 31 (1975) 99–107.

W.C. Herndon and M.L. Ellzey Jr., The construction of isospectral graphs, *MATCH Commun. Math. Comput. Chem.* 20 (1986) 53–79.

A. Hinchliffe, *Chemical Modelling: Application and Theory,* Royal Society of Chemistry, Vol. 3, Cambridge, UK, 2004.

H. Hosoya, Topological index. A newly proposed quantity characterizing the topological nature of structural isomers of saturated hydrocarbons, *Bull. Chem. Soc. Jpn.* 44 (1971) 2332–2339.

H. Hosoya, Mathematical foundation of the organic electron theory—How do $\pi$-electrons flow in conjugated systems? *J. Mol. Struct. (Theochem)* 461–462 (1999) 473–482.

H. Hosoya, From how to why. Graph-theoretical verification of quantum-mechanical aspects of $\pi$-electron behaviors in conjugated systems, *Bull. Chem. Soc. Jpn.* 76 (2003) 2233–2252.

Q.-N. Hu, Y.-Z. Liang and K.-T. Fang, The matrix expression, topological index and atomic attribute of molecular topological structure, *J. Data Sci.* 1 (2003) 361–389.

E. Hückel, Quantentheoretische Beiträge zum Benzolproblem. I. Die Elektronenkonfiguration des Benzols und verwandter Verbindungen, *Z. Phys.* 70 (1931) 204–286.

E. Hückel, Quantentheoretische Beiträge zum Benzolproblem. II. Quantentheorie der induzi-erten Polaritäten, *Z. Phys.* 72 (1932a) 310–337.

E. Hückel, Quantentheoretische Beiträge zum Problem der aromatischen und ungsesättigten Verbindugen. III, *Z. Phys.* 76 (1932b) 628–648.

O. Ivanciuc, Design of topological indices. Part 25. Burden molecular matrices and derived structural descriptors for glycine antagonists, *Rev. Roum. Chim.* 46 (2001) 1047–1066.

D. Janežič, A. Miličević, S. Nikolić, and N. Trinajstić, Topological complexity of molecules, in R. Meyers, Ed., *Encyclopedia of Complexity and Systems Science,* Springer, New York, 2009, Vol. 10, pp. 9210–9224.

D. Janežič, A. Miličević, S. Nikolić, N. Trinajstić, and D. Vukičević, Zagreb indices: Extension to weighted graphs representing molecules containing heteroatoms, *Croat. Chem. Acta* 80 (2007) 541–545.

S. Jiang, H. Liang, and F. Bai, New structural parameters and permanents of adjacency matrices of fullerenes, *MATCH Commun. Math. Comput. Chem.* 56 (2006) 131–139.

M. Karelson, *Molecular descriptors in QSAR/QSPR*, Wiley-Interscience, New York, 2000.

D. Kasum, N. Trinajstić, and I. Gutman, Chemical graph theory. III. On the permanental polynomial, *Croat. Chem. Acta* 54 (1981) 321–328.

N. Kezele, L. Klasinc, J.V. Knop, S. Ivaniš, and S. Nikolić, Computing the variable vertex-connectivity index, *Croat. Chem. Acta* 75 (2002) 651–661.

L.B. Kier and L.L. Hall, *Molecular connectivity in chemistry and drug research*, Academic, New York, 1976.

L.B. Kier and L.L. Hall, *Molecular connectivity in structure-activity analysis*, Wiley, New York, 1986.

E.C. Kirby, D.J. Klein, R.B. Mallion, P. Pollak, and H. Sachs, A theorem for counting spanning trees in general chemical graphs and its particular application to toroidal fullerenes, *Croat. Chem. Acta* 77 (2004) 263–278.

G. Kirchhoff, Über die Auflösung der Gleichungen, auf welche man bei der Untersuchung der linearen Verteilung galvanischer Ströme gefürt wird, *Ann. Phys. Chem.* 72 (1847) 497–508; English translation appeared in J.B. O'Toole, On the solution of the equations obtained from the investigation of the linear distribution of galvanic currents, *Trans IRE* CT-5 (1958) 4–7.

D.J. Klein and A. Misra, Minimally Kekulenoid π-networks and reactivity for acyclics, *Croat. Chem. Acta* 77 (2004) 179–191.

D.J. Klein and N. Trinajstić, Hückel rules and electron correlation, *J. Am. Chem. Soc.* 106 (1984) 8050–8056.

J. von Knop, I. Gutman, and N. Trinajstić, Application of graphs in chemistry. VII. The representation of chemical structures in documentation, *Kem. Ind.* (Zagreb) 24 (1975) 25–29.

J. von Knop, W.R. Müller, K. Szymanski, and N. Trinajstić, *Computer generation of certain classes of molecules*, Kemija u Industriji/SKTH, Zagreb, 1985.

J. von Knop, K. Szymanski, W.R. Müller, H.W. Kroto, and N. Trinajstić, Computer enumeration and generation of physical trees, *J. Comput. Chem.* 8 (1987) 549–554.

J. Konc, M. Hodošček, and D. Janežič, Molecular surface walk, *Croat. Chem. Acta* 79 (2006) 237–241.

D. König, *Theorie der endlichen und und unendlichen Graphen*, Akademische Verlagsgesellschaft, Leipzig, 1936, p. 170.

M. Kunz, A Möbius inversion of the Ulam subgraphs conjecture, *J. Math. Chem.* 9 (1992) 297–305.

M. Kunz, On topological and geometrical distance matrices, *J. Math. Chem.* 13 (1993) 145 151.

M. Kunz, An equivalence relation between distance and coordinate matrices, *MATCH Commun. Math. Comput. Chem.* 32 (1995) 193–203.

X. Li and I. Gutman, *Mathematical aspects of Randić-type molecular structure descriptors*, *Mathematical Chemistry Monographs, MCM-1*, University of Kragujevac and Faculty of Science, Kragujevac, Serbia, 2006.

B. Lučić, S. Nikolić, and N. Trinajstić, Zagreb indices, in *Chemical Information and Computational Challenges in the 21st Century—A Celebration of 2011 International Year of Chemistry*, ed. M.V. Putz, Nova, New York, 2011, pp. 261–275.

B. Lučić, S. Nikolić, N. Trinajstić, B. Zhou, and S. Ivaniš Turk, Sum-connectivity index, in *Novel molecular structure descriptors—Theory and application I*, ed. I. Gutman and B. Furtula, University of Kragujevac, Kragujevac, Serbia, 2010, pp. 101–136.

B. Lučić, I. Sović, J. Batista, K. Skala, D. Plavšić, D. Vikić-Topić, D. Bešlo, S. Nikolić, and N. Trinajstić, The sum-connectivity index—An additive variant of the Randić connectivity index, *Curr. Comput. Aided Drug Des.* 9 (2013) 184–194.

B. Lučić, I. Sović, and N. Trinajstić, The four connectivity matrices, their indices, polynomials and spectra, *Adv. Math. Chem. Appl.* 1 (2014) 76–79.

B. Lučić, N. Trinajstić, S. Sild, M. Karelson, and A.R. Katritzky, A new efficient approach for variable selection based on multiregression: Prediction of gas chromatographic retention times and response factors, *J. Chem. Inf. Comput Sci.* 39 (1999) 610–621.

B. Lučić, N. Trinajstić, and B. Zhou, Comparison between the sum-connectivity and product-connectivity indices for benzenoid hydrocarbons, *Chem. Phys. Lett.* 475 (2009) 146–148.

I. Lukovits, A compact form of the adjacency matrix, *J. Chem. Inf. Comput. Sci.* 40 (2000) 1147–1150.

I. Lukovits, The generation of formulas for isomers, in *Topology in chemistry: Discrete mathematics of molecules*, ed. D.H. Rouvray and R.B. King, Horwood, Chichester, 2002, pp. 327–337.

I. Lukovits, Constructive enumeration of chiral isomers of alkanes, *Croat. Chem. Acta* 77 (2004) 295–300.

I. Lukovits and I. Gutman, On Morgan trees, *Croat. Chem. Acta* 75 (2002) 563–576.

I. Lukovits, A. Miličević, S. Nikolić, and N. Trinajstić, On walk counts and complexity of general graphs, *Internet Electronic J. Mol. Des.* 1 (2002) 388–400. http://www.biochempress.com

I. Lukovits and N. Trinajstić, Atomic walk counts of negative order, *J. Chem. Inf. Comput. Sci.* 43 (2003) 1110–1114.

R.B. Mallion, Some graph-theoretical aspects of simple ring current calculations on conjugated systems, *Proc. Roy. Soc. (London) A* 341 (1975) 429–449.

R.B. Mallion, A.J. Schwenk, and N. Trinajstić, A graphical study of heteroconjugated molecules, *Croat. Chem. Acta* 46 (1974a) 171–182.

R.B. Mallion, A.J. Schwenk, and N. Trinajstić, On the characteristic polynomial of a rooted graph, in *Recent advances in graph theory*, ed. M. Fiedler, Academia, Prague, 1975, pp. 345–350.

R.B. Mallion and N. Trinajstić, Reciprocal spanning-tree density: A new index characterising the intricacy of (poly)cyclic molecular-graph, *MATCH Commun. Math. Comput. Chem.* 48 (2003) 97–116.

R.B. Mallion, N. Trinajstić, and A.J. Schwenk, Graph theory in chemistry. Generalization of Sachs' formula, *Z. Naturforsch.* 29a (1974b) 1481–1484.

R.A. Marcus, Additivity of heats of combustion, LCAO resonance energies and bond orders of conformal sets of conjugated compounds, *J. Chem. Phys.* 43 (1963) 2643–2654.

S. Marković, I. Gutman, and Ž. Bančević, Correlation between Wiener and quasi-Wiener indices in benzenoid hydrocarbons, *J. Serb. Chem. Soc.* 60 (1995) 633–636.

E. Meyer, The IDC system for chemical documentation, *J. Chem. Doc.* 9 (1969) 109–113.

A. Miličević and S. Nikolić, On variable Zagreb indices, *Croat. Chem. Acta* 77 (2004) 97–101.

A. Miličević, S. Nikolić, and N. Trinajstić, On reformulated Zagreb indices, *Mol. Diversity* 8 (2004) 393–399.

H. Minc, *Permanents*, Addison-Wesley, Reading, MA, 1978.

B. Mohar, Laplacian matrices of graphs, in *MATH/CHEM/COMP 1988*, ed. A. Graovac, Elsevier, Amsterdam, 1989, pp. 1–8.

B. Mohar, Eigenvalues, diameter and mean distance in graphs, *Graphs Comb.* 7 (1991) 53–64.

B. Mohar, D. Babić, and N. Trinajstić, A novel definition of the Wiener index for trees, *J. Chem. Inf. Comput. Sci.* 33 (1993) 153–154.

J.W. Moon, *Counting labelled trees*, Vol. 1, Canadian Mathematical Monographs, Ottawa, 1970, chap. 5.

H. Narumi, New topological indices for finite and infinite systems, *MATCH Commun. Math. Comput. Chem.* 22 (1987) 195–207.

S. Nikolić, G. Kovačević, A. Miličević, and N. Trinajstić, The Zagreb indices 30 years after, *Croat. Chem. Acta* 76 (2003a) 113–124.

S. Nikolić, A. Miličević, N. Trinajstić and A. Jurić, On the use of the variable Zagreb $^vM_2$ index in QSPR: Boiling points of benzenoid hydrocarbons, *Molecules* 9 (2004) 1208–1221.

S. Nikolić, I.M. Tolić, N. Trinajstić, and I. Baučić, On the Zagreb indices as complexity indices, *Croat. Chem. Acta* 73 (2000a) 909–921.

S. Nikolić and N. Trinajstić, Compact molecular codes for annulenes, aza-annulenes, annulenoannulenes, aza-annulenoannulenes, cyclazines and aza-cyclazines, *Croat. Chem. Acta* 63 (1990) 155–169.

S. Nikolić and N. Trinajstić, Complexity of molecules, in *Proceedings of the International Conference of Computational Methods in Sciences and Engineering*, ed. T.E. Simos, World Scientific, Singapore, 2003, pp. 454–456.

S. Nikolić, N. Trinajstić, D. Amić, D. Bešlo, and S.C. Basak, Modeling the solubility of aliphatic alcohols in water. Graph connectivity indices vs. line graph connectivity indices, in *QSPR/QSAR Studies by Molecular Descriptors*, ed. M.V. Dudea, Nova, Huntington, NY, 2001, 63–81.

S. Nikolić, N. Trinajstić, and I. Baučić, Comparison between the vertex- and edge-connectivity indices for benzenoid hydrocarbons, *J. Chem. Inf. Comput. Sci.* 38 (1998) 42–46.

S. Nikolić, N. Trinajstić, and S. Ivaniš, The connectivity indices of regular graphs, *Croat. Chem. Acta* 72 (1999a) 875–883.

S. Nikolić, N. Trinajstić, and S. Ivaniš Turk, On the additive version of the connectivity indeks, in T.E. Simos and G. Maroulis, Eds., International Conference of Computational Methods in Science and Engineering, *AIP Conf. Proc.* 1504 (2012) 342–350.

S. Nikolić, N. Trinajstić, A. Jurić, Z. Mihalić, and G. Krilov, Complexity of some interesting (chemical) graphs, *Croat. Chem. Acta* 69 (1996) 883–897.

S. Nikolić, N. Trinajstić, J. von Knop, W.R. Müller, and K. Szymanski, On the concept of the weighted spanning tree of dualist, *J. Math. Chem.* 4 (1990) 357–375.

S. Nikolić, N. Trinajstić, and I.M. Tolić, On the complexity of molecular graphs, *MATCH Commun. Math. Comput. Chem.* 40 (1999b) 187–201.

S. Nikolić, N. Trinajstić, and I.M. Tolić, Complexity of molecules, *J. Chem. Inf. Comput. Sci.* 40 (2000b) 920–926.

S. Nikolić, N. Trinajstić, I.M. Tolić, G. Rücker, and C. Rücker, On molecular complexity indices, in *Complexity in chemistry—Introduction and fundamentals*, ed. D. Bonchev and D.H. Rouvray, Taylor and Francis, London, 2003b, pp. 29–76.

X.-L. Peng, K.-T. Fang, Q.-N. Hu, and Y.-Z. Liang, Impersonality of the connectivity index and recomposition of topological indices according to different properties, *Molecules* 9 (2004) 1089–1099.

J.K. Percus, One more technique for the dimer problem, *J. Math. Phys.* 10 (1969) 1881–1884.

J.K. Percus, *Combinatorial methods*, Springer, Berlin, 1971.

T. Piližota, B. Lučić, and N. Trinajstić, Use of variable selection in modeling the secondary structural content of proteins from their composition of amino acid residues, *J. Chem. Inf. Comput. Sci.* 44 (2004) 113–121.

J.R. Platt, Influence of neighbor bonds on additive bond properties in paraffins, *J. Chem. Phys.* 15 (1947) 419–420.

L. Pogliani, From molecular connectivity to semiempirical terms: Recent trends in graph-theoretical descriptors, *Chem. Rev.* 100 (2000) 3827–3858.

L. Pogliani, Higher-level descriptors in molecular connectivity, *Croat. Chem. Acta* 75 (2002) 409–432.

M. Randić, On the recognition of identical graphs representing molecular topology, *J. Chem. Phys.* 60 (1974) 3920–3928.

M. Randić, On characterization of molecular branching, *J. Am. Chem. Soc.* 97 (1975) 6609–6615.

M. Randić, Random walks and their diagnostic value for characterization of atomic environment, *J. Comput. Chem.* 4 (1980) 386–399.

M. Randić, On molecular identification numbers, *J. Chem. Inf. Comput. Sci.* 24 (1984a) 164–175.

M. Randić, Nonempirical approach to structure-activity studies, *Int. J. Quantum Chem. Quantum Biol. Symp.* 11 (1984b) 137–153.

M. Randić, Compact molecular codes, *J. Chem. Inf. Comput. Sci.* 26 (1986) 136–148.

M. Randić, Novel graph theoretical approach to heterosystems in quantitative structure-activity relationships, *Chemom. Intell. Lab. Syst.* 10 (1991a) 213–227.

M. Randić, On computation of optimal parameters for multivariate analysis of structure-property relationship, *J. Comput. Chem.* 12 (1991b) 970–980.

M. Randić, Similarity based on extended basis descriptors, *J. Chem. Inf. Comput. Sci.* 32 (1992) 686–692.

M. Randić, Complexity of transitive graphs representing degenerate rearrangements, *Croat. Chem. Acta* 74 (2001a) 683–705.

M. Randić, The connectivity index 25 years after, *J. Mol. Graphics Modell.* 20 (2001b) 19–35.

M. Randić, Very efficient search for protein alignment—VESPA, *J. Comput. Chem.* 33 (2012) 702–707.

M. Randić and S.C. Basak, On the use of the variable connectivity index $^1\chi^f$ in QSAR: Toxicity of aliphatic ethers, *J. Chem. Inf. Comput. Sci.* 41 (2001) 614–618.

M. Randić, M. Barysz, J. Nowakowski, S. Nikolić, and N. Trinajstić, Isospectral graphs revisited, *J. Mol. Struct. (Theochem)* 185 (1989) 95–121.

M. Randić, V. Katović, and N. Trinajstić, Symmetry properties of chemical graphs. VII. Enantiomers of a tetragonal-pyramidal rearrangement, in *Symmetries and properties of non-rigid molecules: A comprehensive survey*, ed. J. Maruani and J. Serre, Elsevier, Amsterdam, 1983, pp. 399–408.

M. Randić, S. Nikolić, and N. Trinajstić, Compact molecular codes for polycyclic systems, *J. Mol. Struct. (Theochem)* 165 (1988) 213–228.

M. Randić, T. Pisanski, M. Novič, and D. Plavšić, Novel graph distance matrix, *J. Comput. Chem.* 31 (2010) 1832–1841.

M. Randić and D. Plavšić, On the concept of molecular complexity, *Croat. Chem. Acta* 75 (2002) 107–116.

M. Randić and D. Plavšić, Characterization of molecular complexity, *Int. J. Quntum Chem.* 91 (2003) 20–31.

M. Randić, D. Plavšić, and N. Lerš, Variable connectivity index for cycle-containing structures, *J. Chem. Inf. Comput. Sci.* 41 (2001) 657–662.

M. Randić, M. Pompe, D. Mills, and S.C. Basak, Variable connectivity index as a tool for modeling structure-property relationships, *Molecules* 9 (2004) 1177–1193.

M. Randić and N. Trinajstić, Notes on some less known early contributions to chemical graph theory, *Croat. Chem. Acta* 67 (1994) 1–35.

M. Randić, N. Trinajstić, and T. Živković, Molecular graphs having identical spectra, *J. Chem. Soc. Farady Trans. II* (1976) 244–256.

L.C. Ray and R.A. Kirsch, Finding chemical records by digital computers, *Science* 126 (1957) 814–819.

M. Razinger, Discrimination and ordering of chemical structures by the number of walks, *Theoret. Chim. Acta* 70 (1986) 365–378.

R.C. Read, The enumeration of acyclic chemical compounds, in *Chemical applications of graph theory*, ed. A.T. Balaban, Academic, London, 1976, pp. 25–61.

F.S. Roberts, Graph theory and the social sciences, in *Applications of graph theory*, ed. R.J. Wilson and L.W. Beineke, Academic, London, 1979, pp. 255–291.

D.H. Rouvray, The topological matrix in quantum chemistry, in *Chemical applications of graph theory*, ed. A.T. Balaban, Academic, London, 1976, pp. 175–221.

G. Rücker and C. Rücker, On using the adjacency matrix power method for perception of symmetry and for isomorphism testing of highly intricate graphs, *J. Chem. Inf. Comput. Sci.* 31 (1991) 123–126.

G. Rücker and C. Rücker, Counts of all walks as atomic and molecular descriptors, *J. Chem. Inf. Comput. Sci.* 33 (1993) 683–695.

G. Rücker and C. Rücker, On topological indices, boiling points and cycloalkanes, *J. Chem. Inf. Comput. Sci.* 39 (1999) 788–802.

G. Rücker and C. Rücker, Walk counts, labyrinthicity and complexity of acyclic and cyclic graphs and molecules, *J. Chem. Inf. Comput. Sci.* 40 (2000) 99–106.

G. Rücker and C. Rücker, Substructure, subgraph and walk counts as measures of the complexity of graphs and molecules, *J. Chem. Inf. Comput. Sci.* 41 (2001) 1457–1462.

G. Rücker and C. Rücker, Walking backward: Walk counts of negative order, *J. Chem. Inf. Comput. Sci.* 43 (2003) 1115–1120.

H.P. Schultz, E.B. Schultz, and T.P. Schultz, Topological organic chemistry. 4. Graph theory, matrix permanents and topological indices of alkanes, *J. Chem. Inf. Comput. Sci.* 32 (1992) 69–72.

R.P. Sheridan, The most common chemical replacements in drug-like compounds, *J. Chem. Inf. Comput. Sci.* 42 (2002) 103–108.

L. Spialter, The atom connectivity matrix (ACM) and its characteristic polynomial (ACMCP): A new computer-oriented chemical nomenclature, *J. Am. Chem. Soc.* 85 (1963) 2012–2013.

L. Spialter, The atom connectivity matrix (ACM) and its characteristic polynomial (ACMCP), *J. Chem. Doc.* 4 (1964a) 261–269.

L. Spialter, The atom connectivity matrix characteristic polynomial (ACMCP) and its physico-geometric (topological) significance, *J. Chem. Doc.* 4 (1964b) 269–274.

A. Streitwieser Jr., *Molecular orbital theory for organic chemists*, Wiley, New York, 1961.

K. Szymanski, W.R. Müller, J. von Knop, and N. Trinajstić, On Randić's molecular identification numbers, *J. Chem. Inf. Comput. Sci.* 25 (1985) 413–415.

O.N. Temkin, A.V. Zigarnik, and D. Bonchev, *Chemical reaction networks: A graph-theoretical approach*, CRC, Boca Raton, FL, 1996.

R. Todeschini and V. Consonni, *Handbook of molecular descriptors*, Wiley-VCH, Weinheim, 2000.

R. Todeschini and V. Consonni, *Molecular descriptors for chemoinformatics*, Vols. I and II, Wiley-VCH, Weinheim, 2009.

F. Torrens, Computing the Kekulé structure count for alternant hydrocarbons, *Int. J. Quantum Chem.* 88 (2002) 392–397.

N. Trinajstić, New developments in Hückel theory, *Int. J. Quantum Chem.* S 11 (1977a) 469–477.

N. Trinajstić, Hückel theory and topology, in *Semiempirical methods of electronic structure calculations—Part A: Techniques*, ed. G.J. Segal, Plenum Press, New York, 1977b, pp. 1–27.

N. Trinajstić, *Chemical graph theory*, Vols. I and II, CRC, Boca Raton, FL, 1983.

N. Trinajstić, The role of graph theory in chemistry, *Reports in Molecular Theory* 1 (1990) 185–213.

N. Trinajstić, *Chemical graph theory*, 2nd ed., CRC, Boca Raton, FL, 1992.

N. Trinajstić, D. Babić, S. Nikolić, D. Plavšić, D. Amić, and Z. Mihalić, The Laplacian matrix in chemistry, *J. Chem. Inf. Comput. Sci.* 34 (1994) 368–376.

N. Trinajstić, S. Nikolić, D. Babić, and Z. Mihalić, The vertex- and edge-connectivity indices of Platonic and Archimedean molecules, *Bull. Chem. Technol. Macedonia* 16 (1997) 43–51.

N. Trinajstić, S. Nikolić, J. von Knop, W.R. Müller, and K. Szymanski, *Computational chemical graph theory—Characterization, enumeration and generation of chemical structures by computer methods*, Horwood/Simon & Schuster, New York, 1991, pp. 263–266.

N. Trinajstić, S. Nikolić, A. Miličević, I. Gutman, O. Zagrebačkim indeksima (About Zagreb indices). *Kem. Ind.* 59 (2010) 577–589.

N. Trinajstić and T. Toth, Interactions between users and the secondary sources of information: A case study of Chemical Abstract Services, *Kem. Ind.* (Zagreb) 35 (1986) 527–550.

I. Ugi, J. Bauer, J. Brandt, J. Friedrich, J. Gasteiger, C. Jochum, and W. Schubert, New applications of computers in chemistry, *Angew. Chem. Int. Ed. Engl.* 18 (1979) 111–123.

I. Ugi, M. Wochner, E. Fontain, J. Bauer, B. Gruber, and R. Kari, Chemical similarity, chemical distance and computer-assisted formalized reasoning by analogy, in *Concepts and applications of molecular similarity*, ed. M.A. Johnson and G.G. Maggiora, Wiley-Interscience, New York, 1990, pp. 239–288.

D. Vukičević, S. Nikolić, and N. Trinajstić, On the path-Zagreb matrix, *J. Math. Chem.* 45 (2009) 538–543.

D. Vukičević and N. Trinajstić, Modified Zagreb $M_2$ index—Comparison with the Randić connectivity index for benzenoid systems, *Croat. Chem. Acta* 76 (2003) 183–187.

D. Vukičević and N. Trinajstić, Bond-additive modeling. 3. Comparison between the product-connectivity index and sum-connectivity index, *Croat. Chem. Acta* 83 (2010) 349–351.

S. Wang, B. Zhou, and N. Trinajstić, On the sum-connectivity index, *Filomat* 25 (2011) 29–42.

G.W. Wheland, Syllabus for Advanced Organic Chemistry 321, University of Chicago, Chicago, 1946; see also G.W. Wheland, *Advanced organic chemistry*, Wiley, New York, 1949.

R. Xing, B. Zhou, and N. Trinajstić, Sum-connectivity index of molecular trees, *J. Math. Chem.* 48 (2010) 583–591.

C. Yang and C. Zhong, Modified connectivity indices and their application to QSPR study, *J. Chem. Inf. Comput. Sci.* 43 (2003) 1998–2004.

Y.-Q. Yang, L. Xu, and C.-Y. Hu, Extended adjacency matrix indices and their applications, *J. Chem. Inf. Comput. Sci.* 34 (1994) 1140–1145.

B. Zhou, Zagreb indices, *MATCH Commun. Math. Comput. Chem.* 52 (2004) 113–118.

B. Zhou and I. Gutman, Further properties of Zagreb indices, *MATCH Commun. Math. Comput. Chem.* 54 (2005) 233–239.

B. Zhou and N. Trinajstić, On extended connectivity indices, *J. Math. Chem.* 46 (2009a) 1172–1180.

B. Zhou and N. Trinajstić, On a novel connectivity index, *J. Math. Chem.* 46 (2009b) 1252–1270.

B. Zhou and N. Trinajstić, On sum-connectivity matrix and sum-connectivity energy of a (molecular) graph, *Acta Chim. Slov.* 57 (2010a) 518–523.

B. Zhou and N. Trinajstić, On general sum-connectivity index, *J. Math. Chem.* 47 (2010b) 210–218.

B. Zhou and N. Trinajstić, Mathematical properties of molecular descriptors based on distances, *Croat. Chem. Acta* 83 (2010c) 227–242.

B. Zhou and N. Trinajstić, Minimum general sum-connectivity index of unicyclic graphs, *J. Math. Chem.* 48 (2010d) 697–703.

B. Zhou and N. Trinajstić, Relations between the product- and sum-connectivity indices, *Croat. Chem. Acta* 85 (2012) 363–365.

T. Živković, M. Randić, and N. Trinajstić, On conjugated molecules with identical topological spectra, *Mol. Phys.* 30 (1975) 517–532.

# 3 Incidence Matrices

Any graph is completely determined by either its *adjacencies* or its *incidences*. This can be restated as: graph adjacencies lead to the *adjacency matrices* and graph incidences to the *incidence matrices*, respectively. While the (vertex-)adjacency matrix and its properties have been studied rather thoroughly (e.g., Cvetković et al., 1988, 1995), the (vertex-edge) incidence matrix has been studied less (Rouvray, 1976), although it appears the *vertex-edge* and *edge-cycle* incidence matrices were introduced earlier than the adjacency matrix. For example, at the turn of the century, Poincaré (1900) emphasized these matrices when he presented essentially equivalent *tableaux* appearing in an approach for the construction of geometrical objects (called *complexes* following Listing (1861)) from elementary units, called *cells*. In order to describe how the cells fit together, Poincaré used the Kirchhoff technique (Kirchhoff, 1847), replacing a system of linear equations by a matrix that he built from his considerations of 0-cells and 1-cells. In present-day terminology the *0-cells* and *1-cells* are called *vertices* and *edges*, which together form a *graph*. The corresponding matrix is now known as the *vertex-edge incidence matrix*. Notably, on the strength of this and related papers, Poincaré is regarded as a founder of algebraic topology (Biggs et al., 1976).

## 3.1 THE VERTEX-EDGE INCIDENCE MATRIX

The *vertex-edge incidence matrix* of a graph $G$, denoted by **VE**, is an unsymmetrical (and, in general, not a square) $V \times E$ matrix, which is determined by the incidences of vertices and edges in $G$ (Harary, 1971; Johnson and Johnson, 1972; Bondy and Murty, 1976; Rouvray, 1976; Chartrand, 1977; Trinajstić, 1992; Todeschini and Consonni, 2000, 2009):

$$\left[ \mathbf{VE} \right]_{ij} = \begin{cases} 1 & \text{if the } i\text{-th vertex is incident with the } j\text{-th edge} \\ 0 & \text{otherwise} \end{cases} \tag{3.1}$$

As an example, we give the vertex-edge incidence matrix of a branched tree $T_2$ (Figure 3.1):

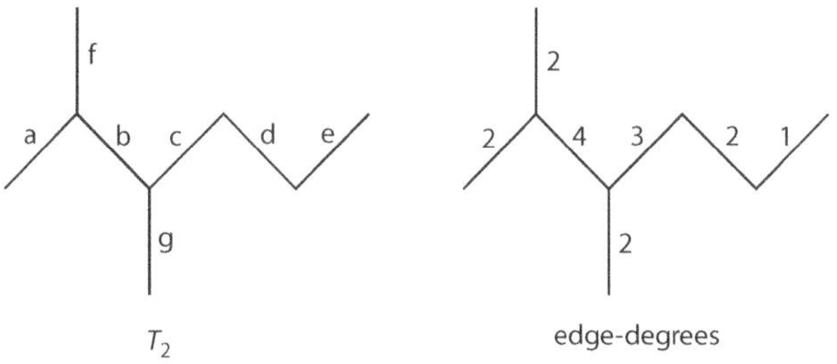

**FIGURE 3.1** Edge-labels and edge-degrees of the branched tree $T_2$ representing the carbon skeleton of 2,3-dimethylhexane. Vertex-labels are given in Figure 2.19.

$$\mathbf{VE}(T_2) = \begin{array}{c} \\ 1 \\ 2 \\ 3 \\ 4 \\ 5 \\ 6 \\ 7 \\ 8 \end{array} \begin{array}{ccccccc} a & b & c & d & e & f & g \\ \left[ \begin{array}{ccccccc} 1 & 0 & 0 & 0 & 0 & 0 & 0 \\ 1 & 1 & 0 & 0 & 0 & 1 & 0 \\ 0 & 1 & 1 & 0 & 0 & 0 & 1 \\ 0 & 0 & 1 & 1 & 0 & 0 & 0 \\ 0 & 0 & 0 & 1 & 1 & 0 & 0 \\ 0 & 0 & 0 & 0 & 1 & 0 & 0 \\ 0 & 0 & 0 & 0 & 0 & 0 & 1 \\ 0 & 0 & 0 & 0 & 0 & 1 & 0 \end{array} \right] \end{array}$$

The above definition (3.1) may be further extended to *oriented graphs*, graphs in which all edges have an assigned direction. If a graph $G$ is oriented, then the nonzero elements of the vertex-edge incidence matrix of $G$ are either +1 or –1, depending on the direction of the edges (Johnson and Johnson, 1972). The +1 values indicate positively incident edges, while the –1 values indicate negatively incident edges (Bryant, 1979). An example of an oriented polycyclic graph is given in Figure 3.2.

The vertex-edge incident matrix associated with $G_{14}$ is given below:

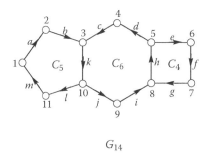

$G_{14}$

**FIGURE 3.2** Oriented tricyclic graph.

$$
\mathbf{VE}(G_{14}) = 
\begin{array}{c}
\phantom{0}\\1\\2\\3\\4\\5\\6\\7\\8\\9\\10\\11
\end{array}
\begin{array}{ccccccccccccc}
a & b & c & d & e & f & g & h & i & j & k & l & m \\
\end{array}
\left[
\begin{array}{ccccccccccccc}
1 & 0 & 0 & 0 & 0 & 0 & 0 & 0 & 0 & 0 & 0 & 0 & -1 \\
-1 & 1 & 0 & 0 & 0 & 0 & 0 & 0 & 0 & 0 & 0 & 0 & 0 \\
0 & -1 & -1 & 0 & 0 & 0 & 0 & 0 & 0 & 0 & 1 & 0 & 0 \\
0 & 0 & 1 & -1 & 0 & 0 & 0 & 0 & 0 & 0 & 0 & 0 & 0 \\
0 & 0 & 0 & 1 & 1 & 0 & 0 & -1 & 0 & 0 & 0 & 0 & 0 \\
0 & 0 & 0 & 0 & -1 & 1 & 0 & 0 & 0 & 0 & 0 & 0 & 0 \\
0 & 0 & 0 & 0 & 0 & -1 & 1 & 0 & 0 & 0 & 0 & 0 & 0 \\
0 & 0 & 0 & 0 & 0 & 0 & 1 & 1 & 1 & 0 & 0 & 0 & 0 \\
0 & 0 & 0 & 0 & 0 & 0 & 0 & 0 & 1 & -1 & 0 & 0 & 0 \\
0 & 0 & 0 & 0 & 0 & 0 & 0 & 0 & 0 & 1 & -1 & 1 & 0 \\
0 & 0 & 0 & 0 & 0 & 0 & 0 & 0 & 0 & 0 & 0 & -1 & 1 \\
\end{array}
\right]
$$

A more general description of the vertex-edge incidence matrix can be given. A vertex-edge incidence matrix $\mathbf{VE}$ with rows and columns labeled by members of two sets $I$ (with the members denoted by $i$) and $J$ (with the members denoted by $j$) of subgraphs of a graph $G$ can be defined as

$$
[\mathbf{VE}]_{ij} = 
\begin{cases}
1 & \text{if sets } I \text{ and } J \text{ have a non-zero intersection} \\
0 & \text{otherwise}
\end{cases}
\tag{3.2}
$$

If $I$ is the set of vertices and $J$ the set of edges, (3.2) reduces to (3.1).

The vertex-edge incidence matrix found some use in chemistry; e.g., Balandin in 1940 employed this matrix, called the *property matrix*, in his study of the physical and chemical properties of molecules (Balandin, 1940), though this work seems to have been largely overlooked (Randić and Trinajstić, 1994). More recently, a few information-theoretic indices have been based on this matrix (Bonchev and Trinajstić, 1977; Bonchev, 1983; Magnuson et al., 1983; Todeschini and Consonni, 2000, 2009).

## 3.2 THE EDGE-VERTEX INCIDENCE MATRIX

The *edge-vertex incidence matrix* $\mathbf{EV}$ is an unsymmetrical $E \times V$ matrix, which is the transpose of the vertex-edge incidence matrix $\mathbf{VE}$. The $\mathbf{EV}$ matrix belonging to $T_2$ (see Figure 3.1) is given below:

$$
\mathbf{EV}(T_2) = 
\begin{array}{c}
\phantom{0}\\a\\b\\c\\d\\e\\f\\g
\end{array}
\begin{array}{cccccccc}
1 & 2 & 3 & 4 & 5 & 6 & 7 & 8 \\
\end{array}
\left[
\begin{array}{cccccccc}
1 & 1 & 0 & 0 & 0 & 0 & 0 & 0 \\
0 & 1 & 1 & 0 & 0 & 0 & 0 & 0 \\
0 & 0 & 1 & 1 & 0 & 0 & 0 & 0 \\
0 & 0 & 0 & 1 & 1 & 0 & 0 & 0 \\
0 & 0 & 0 & 0 & 1 & 1 & 0 & 0 \\
0 & 1 & 0 & 0 & 0 & 0 & 0 & 1 \\
0 & 0 & 1 & 0 & 0 & 0 & 1 & 0 \\
\end{array}
\right]
$$

The incidence matrices **VE** and **EV** are related to the vertex-adjacency matrix $^v\mathbf{A}$ of the graph $G$ as follows (Rouvray, 1976):

$$\mathbf{VE} \times \mathbf{EV} = {}^v\mathbf{A}\ (G) + \boldsymbol{\Delta} \tag{3.3}$$

where $\boldsymbol{\Delta}$ is a diagonal matrix.

The matrices **VE** and **EV** are also related to the vertex-adjacency matrix $^v\mathbf{A}$ of a line graph $L(G)$ of $G$ (Harary, 1971; Rouvray, 1976; Cvetković et al., 1995):

$$\mathbf{EV} \times \mathbf{VE} = {}^v\mathbf{A}\ (L(G)) - 2\ \mathbf{I} \tag{3.4}$$

where $\mathbf{I}$ is the unit $V \times V$ matrix.

When the incidence matrices **VE** and **EV** are associated with the oriented graph $G$, they are related to the Laplacian matrix $\mathbf{L}$ of $G$:

$$\mathbf{VE} \times \mathbf{EV} = \mathbf{L} \tag{3.5}$$

## 3.3   THE EDGE-CYCLE INCIDENCE MATRIX

The *edge-cycle incidence matrix* of a polycyclic graph $G$, denoted by **EC**, is an $E \times C_n$ matrix ($n$ being the size of the cycle), which is determined by the incidences of edges and cycles in $G$:

$$[\mathbf{EC}]_{ij} = \begin{cases} 1 & \text{if the } i\text{-th edge is incident with the } j\text{-th cycle} \\ 0 & \text{otherwise} \end{cases} \tag{3.6}$$

An example of a simple tricyclic graph is shown in Figure 3.3.

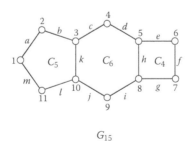

$G_{15}$

**FIGURE 3.3**   A simple tricyclic graph.

The corresponding edge-cycle incidence matrix is given below:

$$
\text{EC}(G_{15}) = 
\begin{array}{c}
 \\
a \\
b \\
c \\
d \\
e \\
f \\
g \\
h \\
i \\
j \\
k \\
l \\
m
\end{array}
\begin{array}{ccc}
C_5 & C_6 & C_4 \\
\left[\begin{array}{ccc}
1 & 0 & 0 \\
1 & 0 & 0 \\
0 & 1 & 0 \\
0 & 1 & 0 \\
0 & 0 & 1 \\
0 & 0 & 1 \\
0 & 0 & 1 \\
0 & 1 & 1 \\
0 & 1 & 0 \\
0 & 1 & 0 \\
1 & 1 & 0 \\
1 & 0 & 0 \\
1 & 0 & 0
\end{array}\right]
\end{array}
$$

## 3.4  THE CYCLE-EDGE INCIDENCE MATRIX

The *cycle-edge incidence matrix* of a polycyclic graph $G$, denoted by **CE**, is a $C_n \times E$ matrix, which is determined by the incidences of cycles and edges in $G$:

$$
[\mathbf{CE}]_{ij} = 
\begin{cases}
1 & \text{if the } i\text{-th cycle is incident with the } j\text{-th edge} \\
0 & \text{otherwise}
\end{cases}
\tag{3.7}
$$

It is evident that the cycle-edge incidence matrix is the *transpose* of the edge-cycle incidence matrix. This matrix is presented because it is used in the counting formula for spanning trees of graphs (see Section 2.20).

The **CE** matrix of $G_{15}$ is given below:

$$
\mathbf{CE}(G_{15}) = 
\begin{array}{c}
C_5 \\
C_6 \\
C_4
\end{array}
\begin{array}{cccccccccccccc}
a & b & c & d & e & f & g & h & i & j & k & l & m \\
\left[\begin{array}{ccccccccccccc}
1 & 1 & 0 & 0 & 0 & 0 & 0 & 0 & 0 & 0 & 1 & 1 & 1 \\
0 & 0 & 1 & 1 & 0 & 0 & 0 & 1 & 1 & 1 & 1 & 0 & 0 \\
0 & 0 & 0 & 0 & 1 & 1 & 1 & 1 & 0 & 0 & 0 & 0 & 0
\end{array}\right]
\end{array}
$$

## 3.5    THE VERTEX-PATH INCIDENCE MATRIX

The *vertex-path incidence matrix*, denoted by **VP**, is a variant of the vertex-edge incidence matrix, where all paths are taken into account instead of only paths of length 1. The generalized definition of incidence matrices, already mentioned above in the case of the **VE** incidence matrix as an expression (3.2), in which the nonvanishing elements of the matrix represent the nonzero intersection of two sets, is convenient to use to set up the **VP** matrices. Therefore, if $V$ is the set of vertices $\{v_i\}$ and $P$ the set of paths $\{p_j\}$ (a *path* being a sequence of edges such that each edge shares one vertex with the sequence-adjacent edges and shares no vertices with any other edge), then the **VP** incidence matrix is defined as

$$[\mathbf{VP}]_{ij} = \begin{cases} n(i,j) & \text{if } \{v_i\} \text{ and } \{p_j\} \text{ have non-zero intersections} \\ 0 & \text{otherwise} \end{cases} \tag{3.8}$$

where $n(i,j)$ is the number of incidences between two sets $\{v_i\}$ and $\{p_j\}$.

Below we give the vertex-path incidence matrix for $T_2$ (see Figure 2.19):

$$\mathbf{VP}(T_2) = \begin{array}{c} \\ v_1 \\ v_2 \\ v_3 \\ v_4 \\ v_5 \\ v_6 \\ v_7 \\ v_8 \end{array} \begin{array}{c} p_0 \quad p_1 \quad p_2 \quad p_3 \quad p_4 \quad p_5 \\ \begin{bmatrix} 1 & 1 & 2 & 2 & 1 & 1 \\ 1 & 3 & 2 & 1 & 1 & 0 \\ 1 & 3 & 3 & 1 & 0 & 0 \\ 1 & 2 & 3 & 2 & 0 & 0 \\ 1 & 2 & 1 & 2 & 2 & 0 \\ 1 & 1 & 1 & 1 & 2 & 2 \\ 1 & 1 & 2 & 3 & 1 & 0 \\ 1 & 1 & 2 & 2 & 1 & 1 \end{bmatrix} \end{array}$$

The vertex-path incidence matrix is also termed the *path-layer matrix* (Skorobogatov and Dobrynin, 1988; Diudea, 1994, 2003), when the paths are organized with respect to length. The **VP** matrix is analogous to the so-called *cardinality layer matrix* (Todeschini and Consonni, 2000, 2009), which has also been called the *path-layer* matrix (Skorobogatov and Dobrynin, 1988), the *distance-frequency* matrix (Diudea and Pârv, 1988), and the *path-sequence matrix* (Todeschini and Consonni, 2000, 2009). Diudea and coworkers (Diudea et al., 1991, 1994; Diudea and Ursu, 2003), Dobrynin (1993), and others (e.g., Hu and Xu, 1996; Yang et al., 2002) derived a number of layer matrices.

## 3.6    THE WEIGHTED-HEXAGON-KEKULÉ-STRUCTURE INCIDENCE MATRIX

Randić (2004) introduced a novel description of Kekulé structures of benzenoids by replacing their standard representation by what he called the *algebraic* representation.

Thus, the represented Kekulé structures Randić named the *algebraic* Kekulé structures (Randić, 2004), while their standard representation he called the *geometric* Kekulé structures. Miličević et al. (2004) and Lučić et al. (2011) used the term *numeric* instead of *algebraic* since the term *numeric* fit better this novel representation of Kekulé structures. The numeric (algebraic) representation of Kekulé structures appears to be useful for linear coding and lexicographic ordering of benzenoids.

The recipe for construction of the numeric Kekulé structures is rather simple: each double bond in the geometric Kekulé structure gets the weight of 2 for two π-electrons making up the double bond, and if the double bond is shared by two hexagons, then it gets the weight of 1. Then, the weight of a hexagon is the sum of contributions from unshared and shared double bonds. Thus, geometric Kekulé structures produce a numeric Kekulé structure that can simply be encoded by a linear sequence of numbers.

The incidence matrix that connects the numeric and geometric Kekulé structures can be constructed using expression (3.1). It is called the *weighted-hexagon-Kekulé-structure incidence matrix* and is denoted by **HK**.

The construction of the weighted-hexagon-Kekulé-structure incidence matrix is exemplified for chrysene. In Figure 3.4, the geometric Kekulé structures of

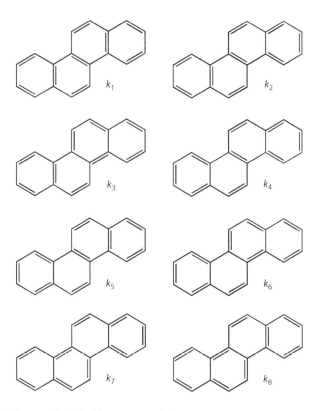

**FIGURE 3.4** Geometric Kekulé structures of chrysene.

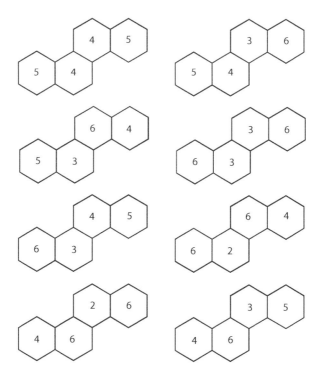

**FIGURE 3.5**   Numeric Kekulé structures of chrysene with the weighted hexagons.

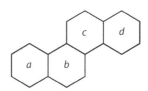

**FIGURE 3.6**   The labeling of hexagons in chrysene.

chrysene are given, and in Figure 3.5, the numeric Kekulé structures of chrysene with the weighted hexagons are presented. In Figure 3.6, we give the labeling of chrysene-hexagons.

Below we give the weighted-hexagon-Kekulé-structure incidence matrix for chrysene:

$$
\begin{array}{c c c c c c c c c}
 & k_1 & k_2 & k_3 & k_4 & k_5 & k_6 & k_7 & k_8 \\
\mathbf{HK}(G) = \begin{array}{c} a \\ b \\ c \\ d \end{array} & \left[\begin{array}{cccccccc}
5 & 5 & 5 & 6 & 6 & 6 & 4 & 4 \\
4 & 4 & 3 & 3 & 3 & 2 & 6 & 6 \\
4 & 3 & 6 & 3 & 4 & 6 & 2 & 3 \\
5 & 6 & 4 & 6 & 5 & 4 & 6 & 5
\end{array}\right]
\end{array}
$$

The concept of numeric Kekulé structures is used for coding and ordering Kekulé structures of *cata*-condensed benzenoids. In the case of *peri*-condensed benzenoids, the numeric code of Kekulé structures is not discriminative enough since there are Kekulé structures with identical numeric codes. In that case, an additional code is needed—Miličević et al. (2004) tested the use of the Wissweser coding system for benzenoid hydrocarbons, and Lučić et al. (2011) tested the perimeter codes.

## REFERENCES

A.A. Balandin, Structral algebra in chemistry, *Acta Physicochim. USSR* 12 (1940) 447–479.

N.L. Biggs, E.K. Lloyd, and R.J. Wilson, *Graph theory 1736–1936*, reprinted with corrections, Clarendon Press, Oxford, 1976; see chap. 8, pp. 131–157. (The latest version is reprinted with corrections in 1998.)

D. Bonchev, *Information theoretic indices for characterization of chemical structures*, Wiley, Chichester, 1983.

D. Bonchev and N. Trinajstić, Information theory, distance matrix and molecular branching, *J. Chem. Phys.* 67 (1977) 4517–4533.

J.A. Bondy and U.S.R. Murty, *Graph theory with applications*, North Holland/Elsevier, Amsterdam, 1976.

P.R. Bryant, Graph theory and electrical networks, in *Applications of graph theory*, ed. R.J. Wilson and L.W. Beineke, Academic, London, 1979, pp. 81–119.

G. Chartrand, *Graphs as mathematical models*, Prindle, Weber and Schmidt, Boston, 1977.

D. Cvetković, M. Doob, I. Gutman, and A. Torgašev, *Recent results in the theory of graph spectra*, North-Holland, Amsterdam, 1988.

D. Cvetković, M. Doob, and H. Sachs, *Spectra of graphs—Theory and applications*, 3rd ed., Johann Ambrosius Barth Verlag, Heidelberg, 1995.

M.V. Diudea, Molecular topology. 16. Layer matrix in molecular graphs, *J. Chem. Inf. Comput. Sci.* 34 (1994) 1064–1071.

M.V. Diudea, Layer matrices and distance property descriptors, *Ind. J. Chem.* 41A (2003) 1283–1294.

M.V. Diudea, O.M. Minailiuc, and A.T. Balaban, Molecular topology. 4. Regressive vertex degrees (new graph invariants) and derived topological indices, *J. Comput. Chem.* 12 (1991) 527–535.

M.V. Diudea and B. Pârv, Molecular topology. 3. A new centric connectivity index, *MATCH Commun. Math. Comput. Chem.* 23 (1988) 65–87.

M.V. Diudea, M. Topan, and A. Graovac, Molecular topology. 17. Layer matrices of walk degrees, *J. Chem. Inf. Comput. Sci.* 34 (1994) 1072–1078.

M.V. Diudea and O. Ursu, Layer matrices and distance property descriptors, *Ind. J. Chem.* 42A (2003) 1283–1294.

A.A. Dobrynin, Degeneracy of some graph invariants, *J. Math. Chem.* 14 (1993) 175–184.

F. Harary, *Graph theory*, 2nd printing, Addison-Wesley, Reading, MA, 1971.

C.-Y. Hu and L. Xu, On highly discriminating molecular topological index, *J. Chem. Inf. Comput. Sci.* 36 (1996) 82–90.

D.E. Johnson and J.R. Johnson, *Graph theory with engineering applications*, Ronald, New York, 1972.

G. Kirchhoff, Über die Auflösung der Gleichungen, auf welche man bei der Untersuchung der linearen Verteilung galvanischer Ströme geführt wird, *Ann. Phys. Chem.* 72 (1847) 497–508; English translation in J.B. O'Toole, On the solution of the equations obtained from the investigation of the linear distribution of galvanic currents, *Trans IRE* CT-5 (1958) 4–7.

J.B. Listing, Der Census räumlicher Complexe, *Nachr. K. Ges. Wiss. Göttingen* (1861) 352–358.

B. Lučić, A. Miličević, S. Nikolić, and N. Trinajstić, Coding and ordering benzenoids and their Kekulé structures, in *Carbon bonding and structures—Advances in physics and chemistry*, ed. M.V. Putz, Springer, New York, 2011, pp. 205–225.

V.R. Magnuson, D.K. Harris, and S.C. Basak, Topological indices based on neighbor symmetry: Chemical and biological applications, in *Chemical applications of topology and graph theory*, ed. R.B. King, Elsevier, Amsterdam, 1983, pp. 178–191.

A. Miličević, S. Nikolić, and N. Trinajstić, Coding and ordering Kekulé structures, *J. Chem. Inf. Comput. Sci.* 44 (2004) 415–421.

H. Poincaré, Second complément à l'Analysis Situs, *Proc. London Math. Soc.* 32 (1900) 277–308.

M. Randić, Algebraic Kekulé structures for benzenoid hydrocarbons, *J. Chem. Inf. Comput. Sci.* 44 (2004) 365–372.

M. Randić and N. Trinajstić, Notes on some less known early contributions to chemical graph theory, *Croat. Chem. Acta* 67 (1994) 1–35.

D.H. Rouvray, The topological matrix in quantum chemistry, in *Chemical applications of graph theory*, ed. A.T. Balaban, Academic, London, 1976, pp. 175–221.

V.A. Skorobogatov and A.A. Dobrynin, Metric analyses of graphs, *MATCH Commun. Math. Comput. Chem.* 23 (1988) 105–151.

R. Todeschini and V. Consonni, *Handbook of molecular descriptors*, Wiley-VCH, Weinheim, 2000.

R. Todeschini and V. Consonni, *Molecular descriptors for chemoinformatics*, Vols. I and II, Wiley-VCH, Weinheim, 2009.

N. Trinjastić, Chemical Graph Theory, 2nd ed., CRC Press, Boca Raton, FL, 1992.

Y. Yang, J. Lin, and C. Wang, Small regular graphs having the same path layer matrix, *J. Graph Theory* 39 (2002) 219–221.

# 4 The Distance Matrix and Related Matrices

Distance matrices are much richer algebraic structures than the adjacency matrices (Harary, 1971; Buckley and Harary, 1990). They are square symmetric $V \times V$ matrices whose entries are graph-theoretical distances between the vertices. Augmented distance matrices have nonzero values on the main diagonal. A number of distance matrices and molecular descriptors derived from them were recently reviewed by Zhou and Trinajstić (2010).

## 4.1 THE STANDARD DISTANCE MATRIX OR THE VERTEX-DISTANCE MATRIX

The *standard distance matrix* or the *vertex-distance matrix* (or the *minimum path matrix*) of a vertex-labeled connected graph $G$ (Harary, 1971; Gutman and Polansky, 1986; Buckley and Harary, 1990; Trinajstić, 1992; Mihalić et al., 1992; Todeschini and Consonni, 2000, 2009; Consonni and Todeschini, 2012), denoted by $^v\mathbf{D}$, is a real symmetric $V \times V$ matrix whose elements are defined as

$$\left[\,^v\mathbf{D}\,\right]_{ij} = \begin{cases} l(i,j) & \text{if } i \neq j \\ 0 & \text{otherwise} \end{cases} \tag{4.1}$$

where $l(i, j)$ is the length of the *shortest* path, i.e., the *minimum* number of edges, between vertices $i$ and $j$ in $G$. The length $l(i, j)$ is also called the *distance* (Harary, 1964, 1967; Hakimi and Yau, 1965; Patrinos and Hakimi, 1973) between vertices $i$ and $j$ in $G$, hence the term *distance matrix*. The term *matrix of lengths* for the distance matrix has also been used (Kruskal, 1956). The shortest distance between two vertices in a graph is also called the *geodesic* distance (Harary, 1971).

The standard distance matrix of the vertex-labeled graph $G_1$ (see structure $A$ in Figure 2.1) is as follows:

$$^v\mathbf{D}(G_1) = \begin{bmatrix} 0 & 1 & 2 & 3 & 4 & 3 & 4 \\ 1 & 0 & 1 & 2 & 3 & 2 & 3 \\ 2 & 1 & 0 & 1 & 2 & 1 & 2 \\ 3 & 2 & 1 & 0 & 1 & 2 & 3 \\ 4 & 3 & 2 & 1 & 0 & 1 & 2 \\ 3 & 2 & 1 & 2 & 1 & 0 & 1 \\ 4 & 3 & 2 & 3 & 2 & 1 & 0 \end{bmatrix}$$

We also give here the vertex-distance matrix of $T_2$ (see Figure 2.19), since we will use this particular matrix later on.

$$
{}^{v}\mathbf{D}(T_2) =
\begin{bmatrix}
0 & 1 & 2 & 3 & 4 & 5 & 3 & 2 \\
1 & 0 & 1 & 2 & 3 & 4 & 2 & 1 \\
2 & 1 & 0 & 1 & 2 & 3 & 1 & 2 \\
3 & 2 & 1 & 0 & 1 & 2 & 2 & 3 \\
4 & 3 & 2 & 1 & 0 & 1 & 3 & 4 \\
5 & 4 & 3 & 2 & 1 & 0 & 4 & 5 \\
3 & 2 & 1 & 2 & 3 & 4 & 0 & 3 \\
2 & 1 & 2 & 3 & 4 & 5 & 3 & 0
\end{bmatrix}
$$

An efficient algorithm is available for computing the vertex-distance matrix of any graph (Müller et al., 1987). This matrix has been used to generate a number of topological indices, e.g., the Wiener index (Wiener, 1947; Hosoya, 1971; Rouvray, 1986; Gutman et al., 1993; Gutman, 1994; Nikolić et al., 1995, 2001b; Diudea, 1996; Dobronyn et al., 2000; Todeschini and Consonni, 2000, 2009; Trinajstić et al., 2001; Bonchev and Klein, 2002; John and Diudea, 2004), modified Wiener indices (Ivanciuc and Klein, 2002; Gutman, 1994; Gutman and Žerovnik, 2002; Gutman et al., 2004), the multiplicative Wiener index (Gutman et al., 2000a, 2000b; Lučić et al., 2001), the hyper-Wiener index (Randić, 1993; Lukovits and Linert, 1994; Klein et al., 1995; Trinajstić et al., 2001; Gutman, 2004), the Balaban index (Balaban, 1983, 1989; Nikolić et al., 2001a; Zhou and Trinajstić, 2009a), and the distance-sum index (Seybold, 1983).

Hosoya (2013) and Hosoya et al. (1994, 2001) observed that two nonisomorphic graphs may possess identical distance-spectra. We already mentioned isospectral graphs when presenting the Hückel matrix (see Section 2.19). A pair of the two polyhedral graphs on eight vertices that possess the same distance-spectra are shown in Figure 4.1.

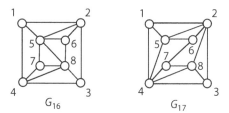

**FIGURE 4.1**   A pair of graphs possessing identical distance-spectra. These two polyhedral graphs are Schlegel graphs representing the pair of nonahedra.

Their vertex-distance matrices are different and are given as follows:

$$
{}^v\mathbf{D}(G_{16}) =
\begin{bmatrix}
0 & 1 & 2 & 1 & 1 & 2 & 2 & 2 \\
1 & 0 & 1 & 2 & 1 & 1 & 2 & 2 \\
2 & 1 & 0 & 1 & 2 & 2 & 2 & 1 \\
1 & 2 & 1 & 0 & 2 & 2 & 1 & 1 \\
1 & 1 & 2 & 2 & 0 & 1 & 1 & 1 \\
2 & 1 & 2 & 2 & 1 & 0 & 2 & 1 \\
2 & 2 & 2 & 1 & 1 & 2 & 0 & 1 \\
2 & 2 & 1 & 1 & 1 & 1 & 1 & 0
\end{bmatrix}
$$

$$
{}^v\mathbf{D}(G_{17}) =
\begin{bmatrix}
0 & 1 & 2 & 1 & 1 & 2 & 2 & 2 \\
1 & 0 & 1 & 2 & 1 & 1 & 2 & 1 \\
2 & 1 & 0 & 1 & 2 & 2 & 2 & 1 \\
1 & 2 & 1 & 0 & 1 & 2 & 1 & 1 \\
1 & 1 & 2 & 1 & 0 & 1 & 2 & 2 \\
2 & 1 & 2 & 2 & 1 & 0 & 1 & 2 \\
2 & 2 & 2 & 1 & 2 & 1 & 0 & 1 \\
2 & 1 & 1 & 1 & 2 & 2 & 1 & 0
\end{bmatrix}
$$

The corresponding distance-spectra are identical: $\{G_{16}\} = \{G_{17}\} = \{10.3150, 0.2985, 0.0953, -0.8112, -1.2624, -2.4773, -2.8024, -3.3557\}$.

These kind of graphs are called *twin graphs* (Hosoya et al., 1994, 2001), because they possess, besides identical distance-spectra and consequently identical distance-polynomials, identical characteristic polynomials and their spectra $\{3.8801, 1.3557, 0.7732, 0.4773, -0.7376, -1.2464, -2.0953, -2.4069\}$, identical matching polynomials and their spectra, and many identical graph-theoretical invariants.

The role of the distance matrix in chemistry is scholarly reviewed by Rouvray (1986).

## 4.2 GENERALIZED VERTEX-DISTANCE MATRIX

The generalized vertex-distance matrix, denoted as ${}^v\mathbf{D}^\lambda$, is derived from the distance matrix ${}^v\mathbf{D}$ by raising its elements to an exponent $\lambda$:

$$
\left[{}^v\mathbf{D}^\lambda\right]_{ij} =
\begin{cases}
[l(i,j)]^\lambda & \text{if } i \neq j \\
0 & \text{otherwise}
\end{cases}
\tag{4.2}
$$

where $\lambda$ is any real number. When $\lambda = 1$, ${}^v\mathbf{D}^\lambda$ reduces to the vertex-distance matrix ${}^v\mathbf{D}$.

The squared vertex-distance matrix $^v\mathbf{D}^2$ is given by

$$\left[{}^v\mathbf{D}^2\right]_{ij} = \begin{cases} \left[l(i,j)\right]^2 & \text{if } i \neq j \\ 0 & \text{otherwise} \end{cases} \tag{4.3}$$

while the reciprocal squared vertex-distance matrix $^v\mathbf{D}^{-2}$ is defined as follows:

$$\left[{}^v\mathbf{D}^{-2}\right]_{ij} = \begin{cases} 1/\left[l(i,j)\right]^2 & \text{if } i \neq j \\ 0 & \text{otherwise} \end{cases} \tag{4.4}$$

The reciprocal vertex-distance matrix $^v\mathbf{D}^{-1}$ is called the *Harary matrix* (Plavšić et al., 1993) and is presented in Section 4.24. Below we give the reciprocal squared vertex-distance matrix $^v\mathbf{D}^{-2}$ of $T_2$ (see Figure 2.19).

$$^v\mathbf{D}^{-2}(T_2) = \begin{bmatrix} 0 & 1 & 0.250 & 0.111 & 0.063 & 0.040 & 0.111 & 0.250 \\ 1 & 0 & 1 & 0.250 & 0.111 & 0.063 & 0.250 & 1 \\ 0.250 & 1 & 0 & 1 & 0.250 & 0.111 & 1 & 0.250 \\ 0.111 & 0.250 & 1 & 0 & 1 & 0.250 & 0.250 & 0.111 \\ 0.063 & 0.111 & 0.250 & 1 & 0 & 1 & 0.111 & 0.063 \\ 0.040 & 0.063 & 0.111 & 0.250 & 1 & 0 & 0.063 & 0.040 \\ 0.111 & 0.250 & 1 & 0.250 & 0.111 & 0.063 & 0 & 0.111 \\ 0.250 & 1 & 0.250 & 0.111 & 0.063 & 0.040 & 0.111 & 0 \end{bmatrix}$$

The squared vertex-distance matrix $^v\mathbf{D}^2$ and reciprocal squared vertex-distance matrix $^v\mathbf{D}^{-2}$ have been discussed by Consonni and Todeschini (Todeschini and Consonni, 2000, 2009; Consonni and Todeschini, 2012).

## 4.3 THE VERTEX-GALVEZ MATRIX

The vertex-Galvez matrix of a graph $G$, denoted by $^v\mathbf{GM}$, is a square asymmetric matrix that can be obtained by multiplying the vertex-adjacency matrix $^v\mathbf{A}$ and the reciprocal square vertex-distance matrix $^v\mathbf{D}^{-2}$ (Gálvez et al., 1994, 1995; Todeschini and Consonni, 2000, 2009):

$$^v\mathbf{GM} = {}^v\mathbf{A} \times {}^v\mathbf{D}^{-2} \tag{4.5}$$

As an example, we give the vertex-Galvez matrix for $T_2$ (see Figure 2.19). Below we give the vertex-adjacency matrix of $T_2$, $^v\mathbf{A}(T_2)$, while the reciprocal square vertex-distance matrix $^v\mathbf{D}^{-2}$ is given above.

$$
{}^{v}\mathbf{A}(T_2) =
\begin{bmatrix}
0 & 1 & 0 & 0 & 0 & 0 & 0 & 0 \\
1 & 0 & 1 & 0 & 0 & 0 & 0 & 1 \\
0 & 1 & 0 & 1 & 0 & 0 & 1 & 0 \\
0 & 0 & 1 & 0 & 1 & 0 & 0 & 0 \\
0 & 0 & 0 & 1 & 0 & 1 & 0 & 0 \\
0 & 0 & 0 & 0 & 1 & 0 & 0 & 0 \\
0 & 0 & 1 & 0 & 0 & 0 & 0 & 0 \\
0 & 1 & 0 & 0 & 0 & 0 & 0 & 0
\end{bmatrix}
$$

The $^{v}\mathbf{GM}$ matrix is then given by

$$
{}^{v}\mathbf{GM}(T_2) =
\begin{bmatrix}
1 & 0 & 1 & 0.250 & 0.111 & 0.063 & 0.250 & 1 \\
0.500 & 3 & 0.500 & 1.222 & 0.376 & 0.191 & 1.222 & 0.500 \\
1.111 & 0.500 & 3 & 0.500 & 1.222 & 0.376 & 0.500 & 1.222 \\
0.313 & 1.111 & 0.250 & 2 & 0.250 & 1.111 & 1.111 & 0.313 \\
0.151 & 0.313 & 1.111 & 0.250 & 2 & 0.250 & 0.313 & 0.151 \\
0.063 & 0.111 & 0.250 & 1 & 0 & 1 & 0.111 & 0.063 \\
0.250 & 1 & 0 & 1 & 0.250 & 0.111 & 1 & 0.250 \\
1 & 0 & 1 & 0.250 & 0.111 & 0.063 & 0.250 & 1
\end{bmatrix}
$$

The vertex-Galvez matrix $^{v}\mathbf{GM}$ can be, for example, used to generate the *charge-transfer matrix*. This matrix, denoted as **CTM**, is defined as

$$
[\mathbf{CTM}]_{ij} =
\begin{cases}
m(ij) - m(ji) & \text{if } i \neq j \\
d(i) & \text{if } i = j
\end{cases}
\tag{4.6}
$$

where $d(i)$ is the degree of a vertex $i$ while $m(i, j)$ are elements of the vertex-Galvez matrix $^{v}\mathbf{GM}$.

As an illustrative example, we give below the charge-transfer matrix of $T_2$ (see Figure 2.19):

$$
\mathbf{CTM}(T_2) =
\begin{bmatrix}
1 & -0.500 & -0.111 & -0.063 & -0.040 & 0 & 0 & 0 \\
0.500 & 3 & 0 & 0.111 & 0.063 & 0.080 & 0.222 & 0.500 \\
0.111 & 0 & 3 & 0.250 & 0.111 & 0.126 & 0.500 & 0.222 \\
0.063 & -0.111 & -0.250 & 2 & 0 & 0.111 & 0.111 & 0.063 \\
0.040 & -0.063 & -0.111 & 0 & 2 & 0.250 & 0.063 & 0.040 \\
0 & -0.080 & -0.126 & -0.111 & -0.250 & 1 & 0 & 0 \\
0 & -0.222 & -0.500 & -0.111 & -0.063 & 0 & 1 & 0 \\
0 & -0.500 & -0.222 & -0.063 & -0.040 & 0 & 0 & 0
\end{bmatrix}
$$

The diagonal entries of the charge-transfer matrix $(\mathbf{CTM})_{ii}$ represent the topological valence of the atoms in a molecule, while the off-diagonal elements $(\mathbf{CTM})_{ij}$ represent a measure of the net charge transfer from the atom $j$ to the atom $i$, or if the value of $(\mathbf{CTM})_{ij}$ is negative, the atom $i$ will transfer net charge to atom $j$.

## 4.4  COMBINATORIAL MATRICES

Diudea (1996a, 1996b; Diudea et al., 2006) introduced two *combinatorial matrices*, the *distance-delta matrix* $^\Delta\mathbf{CM}$ and the *distance-path matrix* $^P\mathbf{CM}$, whose entries are based on the elements of the related vertex-distance matrix $^v\mathbf{D}$:

$$\left[^\Delta\mathbf{CM}\right]_{ij} = \begin{cases} \left(\dfrac{\left[^v\mathbf{D}\right]_{ij}}{2}\right) & \text{if } i \neq j \\[2mm] 0 & \text{if } i = j \end{cases} \tag{4.7}$$

$$\left[^P\mathbf{CM}\right]_{ij} = \begin{cases} \left(\dfrac{\left[^v\mathbf{D}\right]_{ij}}{2}\right) & \text{if } i \neq j \\[2mm] 0 & \text{if } i = j \end{cases} \tag{4.8}$$

As examples, we give below $^\Delta\mathbf{CM}$ and $^P\mathbf{CM}$ matrices for the tree $T_2$ (see Figure 2.19) and the graph $G_1$ (see structure $A$ in Figure 2.1). Distance matrices of $T_2$ and $G_1$ are given above.

$$^\Delta\mathbf{CM}(T_2) = \begin{bmatrix} 0 & 0 & 1 & 3 & 6 & 10 & 3 & 1 \\ 0 & 0 & 0 & 1 & 3 & 6 & 1 & 0 \\ 1 & 0 & 0 & 0 & 1 & 3 & 0 & 1 \\ 3 & 1 & 0 & 0 & 0 & 1 & 1 & 3 \\ 6 & 3 & 1 & 0 & 0 & 0 & 3 & 6 \\ 10 & 6 & 3 & 1 & 0 & 0 & 6 & 10 \\ 3 & 1 & 0 & 1 & 3 & 6 & 0 & 3 \\ 1 & 0 & 1 & 3 & 6 & 10 & 3 & 0 \end{bmatrix}$$

$$
{}^{p}\mathbf{CM}(T_2) =
\begin{bmatrix}
0 & 1 & 3 & 6 & 10 & 15 & 6 & 3 \\
1 & 0 & 1 & 3 & 6 & 10 & 3 & 1 \\
3 & 1 & 0 & 1 & 3 & 6 & 1 & 3 \\
6 & 3 & 1 & 0 & 1 & 3 & 3 & 6 \\
10 & 6 & 3 & 1 & 0 & 1 & 6 & 10 \\
15 & 10 & 6 & 3 & 1 & 0 & 10 & 15 \\
6 & 3 & 1 & 3 & 6 & 10 & 0 & 6 \\
3 & 1 & 3 & 6 & 10 & 15 & 6 & 0
\end{bmatrix}
$$

$$
{}^{\Delta}\mathbf{CM}(G_1) =
\begin{bmatrix}
0 & 0 & 1 & 3 & 6 & 3 & 6 \\
0 & 0 & 0 & 1 & 3 & 1 & 3 \\
1 & 0 & 0 & 0 & 1 & 0 & 1 \\
3 & 1 & 0 & 0 & 0 & 1 & 3 \\
6 & 3 & 1 & 0 & 0 & 0 & 1 \\
3 & 1 & 0 & 1 & 0 & 0 & 0 \\
6 & 3 & 1 & 3 & 1 & 0 & 0
\end{bmatrix}
$$

$$
{}^{p}\mathbf{CM}(G_1) =
\begin{bmatrix}
0 & 1 & 3 & 6 & 10 & 6 & 10 \\
1 & 0 & 1 & 3 & 6 & 3 & 6 \\
3 & 1 & 0 & 1 & 3 & 1 & 3 \\
6 & 3 & 1 & 0 & 1 & 3 & 6 \\
10 & 6 & 3 & 1 & 0 & 1 & 3 \\
6 & 3 & 1 & 3 & 1 & 0 & 1 \\
10 & 6 & 3 & 6 & 3 & 1 & 0
\end{bmatrix}
$$

The distance-delta matrix gives a descriptor related to the non-Wiener part of the hyper-Wiener index, while the distance-path matrix allows direct computation of the hyper-Wiener index (Diudea, 1996a, 1996b). It should also be noted that there is the following relationship between the elements of the three matrices based on the shortest graph-distances ${}^{v}\mathbf{D}$, ${}^{p}\mathbf{CM}$, and ${}^{\Delta}\mathbf{CM}$:

$$
[{}^{v}\mathbf{D}]_{ij} = [{}^{p}\mathbf{CM}]_{ij} - [{}^{\Delta}\mathbf{CM}]_{ij} \tag{4.9}
$$

## 4.5   RECIPROCAL COMBINATORIAL MATRICES

The *reciprocal combinatorial matrices*, the reciprocal distance-delta matrix ${}^{\Delta}\mathbf{CM}^{-1}$ and the reciprocal distance-path matrix ${}^{p}\mathbf{CM}^{-1}$, are matrices whose entries are the reciprocals of the corresponding elements in the combinatorial matrices $^\Delta\mathbf{CM}$ and $^P\mathbf{CM}$:

$$\left[{}^\Delta\mathbf{CM}^{-1}\right]_{ij} = \begin{cases} 1/[{}^\Delta\mathbf{CM}]_{ij} & \text{if } i \neq j, \text{ but } i \text{ and } j \text{ are not adjacent} \\ 0 & \text{if } i \text{ and } j \text{ are adjacent} \\ 0 & \text{if } i = j \end{cases} \qquad (4.10)$$

$$\left[{}^P\mathbf{CM}^{-1}\right]_{ij} = \begin{cases} 1/[{}^P\mathbf{CM}]_{ij} & \text{if } i \neq j \\ 0 & \text{if } i = j \end{cases} \qquad (4.11)$$

The reciprocal combinatorial matrices $^\Delta\mathbf{CM}^{-1}$ and $^P\mathbf{CM}^{-1}$ of $T_2$ (see Figure 2.19) and $G_1$ (see structure $A$ in Figure 2.1), based on the corresponding combinatorial matrices, given above, are as follows:

$$^\Delta\mathbf{CM}^{-1}(T_2) = \begin{bmatrix} 0 & 0 & 1 & 1/3 & 1/6 & 1/10 & 1/3 & 1 \\ 0 & 0 & 0 & 1 & 1/3 & 1/6 & 1 & 0 \\ 1 & 0 & 0 & 0 & 1 & 1/3 & 0 & 1 \\ 1/3 & 1 & 0 & 0 & 0 & 1 & 1 & 1/3 \\ 1/6 & 1/3 & 1 & 0 & 0 & 0 & 1/3 & 1/6 \\ 1/10 & 1/6 & 1/3 & 1 & 0 & 0 & 1/6 & 1/10 \\ 1/3 & 1 & 0 & 1 & 1/3 & 1/6 & 0 & 1/3 \\ 1 & 0 & 1 & 1/3 & 1/6 & 1/10 & 1/3 & 0 \end{bmatrix}$$

$$^P\mathbf{CM}^{-1}(T_2) = \begin{bmatrix} 0 & 1 & 1/3 & 1/6 & 1/10 & 1/15 & 1/6 & 1/3 \\ 1 & 0 & 1 & 1/3 & 1/6 & 1/10 & 1/3 & 1 \\ 1/3 & 1 & 0 & 1 & 1/3 & 1/6 & 1 & 1/3 \\ 1/6 & 1/3 & 1 & 0 & 1 & 1/3 & 1/3 & 1/6 \\ 1/10 & 1/6 & 1/3 & 1 & 0 & 1 & 1/6 & 1/10 \\ 1/15 & 1/10 & 1/6 & 1/3 & 1 & 0 & 1/10 & 1/15 \\ 1/6 & 1/3 & 1 & 1/3 & 1/6 & 1/10 & 0 & 1/6 \\ 1/3 & 1 & 1/3 & 1/6 & 1/10 & 1/15 & 1/6 & 0 \end{bmatrix}$$

$$
^{\Delta}\mathbf{CM}(G_1) = \begin{bmatrix}
0 & 0 & 1 & 1/3 & 1/6 & 1/3 & 1/6 \\
0 & 0 & 0 & 1 & 1/3 & 1 & 1/3 \\
1 & 0 & 0 & 0 & 1 & 0 & 1 \\
1/3 & 1 & 0 & 0 & 0 & 1 & 1/3 \\
1/6 & 1/3 & 1 & 0 & 0 & 0 & 1 \\
1/3 & 1 & 0 & 1 & 0 & 0 & 0 \\
1/6 & 1/3 & 1 & 1/3 & 1 & 0 & 0
\end{bmatrix}
$$

$$
^{p}\mathbf{CM}^{-1}(G_1) = \begin{bmatrix}
0 & 1 & 1/3 & 1/6 & 1/10 & 1/6 & 1/10 \\
1 & 0 & 1 & 1/3 & 1/6 & 1/3 & 1/6 \\
1/3 & 1 & 0 & 1 & 1/3 & 1 & 1/3 \\
1/6 & 1/3 & 1 & 0 & 1 & 1/3 & 1/6 \\
1/10 & 1/6 & 1/3 & 1 & 0 & 1 & 1/3 \\
1/6 & 1/3 & 1 & 1/3 & 1 & 0 & 1 \\
1/10 & 1/6 & 1/3 & 1/6 & 1/3 & 1 & 0
\end{bmatrix}
$$

Hyper-Harary indices are defined for the reciprocal distance-path matrix (Diudea, 1997; Todeschini and Consonni, 2000, 2009) and were used in a structure-property modeling of lower alkanes and branched and unbranched cycloalkanes (Trinajstić et al., 2001).

## 4.6 THE EDGE-DISTANCE MATRIX

The *edge-distance matrix* of a graph $G$, denoted by $^{e}\mathbf{D}$, is the vertex-distance matrix of the corresponding line graph $L(G)$:

$$^{e}\mathbf{D}(G) = ^{v}\mathbf{D}[L(G)] \tag{4.12}$$

This is because the edge-distances in a graph are equal to the distances between vertices in the corresponding line graph. We give below the edge-distance matrix of $G_1$ (see structure $B$ in Figure 2.1) as the vertex-distance matrix of $L(G_1)$.

$$
^{e}\mathbf{D}(G_1) = \begin{bmatrix}
0 & 1 & 2 & 3 & 3 & 3 & 2 \\
1 & 0 & 1 & 2 & 2 & 2 & 1 \\
2 & 1 & 0 & 1 & 2 & 2 & 1 \\
3 & 2 & 1 & 0 & 1 & 2 & 2 \\
3 & 2 & 2 & 1 & 0 & 1 & 1 \\
3 & 2 & 2 & 2 & 1 & 0 & 1 \\
2 & 1 & 1 & 2 & 1 & 1 & 0
\end{bmatrix}
$$

The edge-distance matrix of a graph $G$ can also be simply constructed without considering the related line graph by counting vertices between the edges of $G$:

$$\left[{}^{e}\mathbf{D}\right]_{ij} = \begin{cases} n_{ij} + 2 & \text{if } i \neq j \\ 0 & \text{otherwise} \end{cases} \tag{4.13}$$

where $n_{ij}$ is the number of vertices on the shortest path between edges $i$ and $j$.

## 4.7 THE VERTEX-DISTANCE-COMPLEMENT MATRIX

The *vertex-distance-complement matrix* (Balaban et al., 2000; Randić and Zupan, 2001; Nikolić et al., 2001a), denoted by ${}^{vc}\mathbf{D}$, can be obtained simply from the vertex-distance matrix:

$$\left[{}^{vc}\mathbf{D}\right]_{ij} = \begin{cases} V - \left[{}^{v}\mathbf{D}\right]_{ij} & \text{if } i \neq j \\ 0 & \text{otherwise} \end{cases} \tag{4.14}$$

The distance-complement matrix of the vertex-labeled graph $G_1$ (structure $A$ in Figure 2.1) is

$$
{}^{vc}\mathbf{D}(G_1) = \begin{bmatrix}
0 & 6 & 5 & 4 & 3 & 4 & 3 \\
6 & 0 & 6 & 5 & 4 & 5 & 4 \\
5 & 6 & 0 & 6 & 5 & 6 & 5 \\
4 & 5 & 6 & 0 & 6 & 5 & 4 \\
3 & 4 & 5 & 6 & 0 & 6 & 5 \\
4 & 5 & 6 & 5 & 6 & 0 & 6 \\
3 & 4 & 5 & 4 & 5 & 6 & 0
\end{bmatrix}
$$

Several complement distance indices are available (Balaban et al., 2000; Randić and Zupan, 2001; Nikolić et al., 2001a): the complement Wiener index, the complement hyper-Wiener index, the complement Balaban index.

## 4.8 THE AUGMENTED VERTEX-DISTANCE MATRIX

The *augmented vertex-distance matrix* (Randić and Pompe, 2001) of a vertex-labeled connected vertex-weighted graph ${}^{vw}G$, denoted by ${}^{a}\mathbf{D}$, is a real symmetric $V \times V$ matrix whose elements are defined as

$$\left[ {}^a\mathbf{D} \right]_{ij} = \begin{cases} l(i,j) & \text{if } i \neq j \\ \delta_{ii} & \text{if } i = j \end{cases} \tag{4.15}$$

where $\delta_{ii}$ is the variable weight of a vertex $i$.

The augmented distance matrix of the vertex-labeled weighted graph $G_4$ in Figure 2.15 is given by

$$
{}^a\mathbf{D}(G_4) = \begin{bmatrix}
\delta_x & 1 & 2 & 3 & 4 & 3 & 4 \\
1 & \delta_x & 1 & 2 & 3 & 2 & 3 \\
2 & 1 & \delta_x & 1 & 2 & 1 & 2 \\
3 & 2 & 1 & \delta_y & 1 & 2 & 3 \\
4 & 3 & 2 & 1 & \delta_x & 1 & 2 \\
3 & 2 & 1 & 2 & 1 & \delta_x & 1 \\
4 & 3 & 2 & 3 & 2 & 1 & \delta_x
\end{bmatrix}
$$

Here, $\delta_x$ and $\delta_y$ are variable parameters, which are determined during the regression so that the standard error of estimate for a studied property is as small as possible. The augmented distance matrix has been used for generation of a number of variable distance indices (Randić and Pompe, 2001; Lučić et al., 2003): the variable Wiener index, the variable hyper-Wiener index, the variable Balaban index, and variable complements of these indices based on the augmented distance-complement matrix.

## 4.9 THE EDGE-WEIGHTED VERTEX-DISTANCE MATRIX

The *edge-weighted vertex-distance matrix*, denoted by ${}^{ewv}\mathbf{D}$, is a square symmetric $V \times V$ matrix defined as

$$\left[ {}^{ewv}\mathbf{D} \right]_{ij} = \begin{cases} w(p_{ij}) & \text{if } i \neq j \\ 0 & \text{otherwise} \end{cases} \tag{4.16}$$

where $w(p_{ij})$ is the minimum sum of edge-weights along the path between the vertices $i$ and $j$ in an edge-weighted graph $G_{ew}$, which may not be the *shortest* possible path between these two vertices in $G_{ew}$. Hence, in the case of vertex-distance matrix for edge-weighted graphs, the entry $[{}^{ewv}\mathbf{D}]_{ij}$ is the minimum path-weight between the vertices $i$ and $j$. In Figure 4.2, we give the edge-weighted graph $G_{18}$. The edge-weighted vertex-distance matrix of $G_{18}$ is given below the figure.

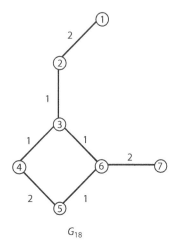

**FIGURE 4.2**   The vertex-labeled edge-weighted graph $G_{18}$. The vertex-labels are encircled.

$$^{ewv}\mathbf{D}(G_{18}) = \begin{bmatrix} 0 & 2 & 3 & 4 & 5 & 4 & 6 \\ 2 & 0 & 1 & 2 & 3 & 2 & 4 \\ 3 & 1 & 0 & 1 & 2 & 1 & 3 \\ 4 & 2 & 1 & 0 & 2 & 2 & 4 \\ 5 & 3 & 2 & 2 & 0 & 1 & 3 \\ 4 & 2 & 1 & 2 & 1 & 0 & 2 \\ 6 & 4 & 3 & 4 & 3 & 2 & 0 \end{bmatrix}$$

The summation of elements in the upper (or the lower) matrix-triangle gives the weighted-Wiener index.

## 4.10   THE BARYSZ VERTEX-DISTANCE MATRIX

The *Barysz vertex-distance matrix* (Todeschini and Consonni, 2000, 2009; Consonni, Todeschini, 2012), denoted by $^{Bv}\mathbf{D}$, is a type of weighted real symmetric $V \times V$ matrix whose entries reflect the vertex- and edge-weights of a weighted graph. As was already stated, the weighted graphs are used to represent heterosystems (Mallion et al., 1974a, 1974b, 1975; Graovac et al., 1975). The Barysz (weighted) vertex-distance matrix is defined as (Barysz et al., 1983)

$$\left[ ^{Bv}\mathbf{D} \right]_{ij} = \begin{cases} w_{ij} & \text{if } i \neq j \\ w_{ii} & \text{if } i = j \text{ and if the vertex } i \text{ weighted} \\ 0 & \text{otherwise} \end{cases} \qquad (4.17)$$

where $w_{ij}$ is the weight of edges between vertices $i$ and $j$ and $w_{ii}$ is the weight of a vertex $i$.

Barysz et al. (1983) defined pragmatically the weight of a vertex $i$ as

$$w_{ii} = 1 - (Z_C/Z_i) \tag{4.18}$$

where $Z_C = 6$ (the atomic number of carbon) and $Z_i$ is the atomic number of the element $i$.

The weight of an edge $i$-$j$ is defined as

$$w_{ij} = \sum_r k_r \tag{4.19}$$

where the parameter $k_r$ is given by

$$k_r = b_r^{-1} (Z_C^2/Z_i Z_j) \tag{4.20}$$

where $b_r$ is the bond multiplicity parameter with values 1, 1.5, 2, and 3 for a single, an aromatic, a double, and a triple bond, respectively. Tables of $w_{ii}$ and $k_r$ parameters are available (Barysz et al., 1983; Trinajstić, 1983, 1992; Randić et al., 1988; Ivanciuc et al., 1999). Below we give the Barysz vertex-distance matrix of $G_4$. The vertex-weight and edge-weights of $G_4$, computed using the Barysz parametrization procedure (Barysz et al., 1983), are given in Figure 4.3. This procedure has also been called the Z-weighting scheme (Ivanciuc et al., 1999).

$$^{Bv}\mathbf{D}(G_4) = \begin{bmatrix} 0 & 1 & 2 & 2.857 & 4 & 3 & 4 \\ 1 & 0 & 1 & 1.857 & 3 & 2 & 3 \\ 2 & 1 & 0 & 0.857 & 2 & 1 & 2 \\ 2.857 & 1.857 & 0.857 & 0.143 & 0.857 & 1.857 & 2.857 \\ 4 & 3 & 2 & 0.857 & 0 & 1 & 2 \\ 3 & 2 & 1 & 1.857 & 1 & 0 & 1 \\ 4 & 3 & 2 & 2.857 & 2 & 1 & 0 \end{bmatrix}$$

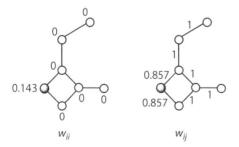

FIGURE 4.3    The vertex-weight ($w_{ii}$) and edge-weights ($w_{ij}$) of $G_4$.

There have also been other equally pragmatic attempts to parameterize the heteroatoms and heterobonds in the distance matrix of weighted graphs (Ivanciuc et al., 1999):

1. The relative electronegativity X-weighting scheme (Balaban, 1986; Balaban and Ivanciuc, 1989): Sanderson electronegativities (Sanderson, 1983).
2. The relative covalent radius Y-weighting scheme (Balaban, 1986): The covalent radius selected from Sanderson (1983).
3. The atomic mass A-weighting scheme (Ivanciuc, 2000a).
4. The AH-scheme (Ivanciuc, 2000a): A weighting scheme based on the atomic mass A that produces parameters different from those of the Z-scheme by considering the mass of heavy (nonhydrogen) atom together with that of hydrogen atoms connected to it.
5. The polarizability P-weighting scheme (Ivanciuc, 2000a).
6. The atomic radius R-weighting scheme (Ivanciuc, 2000a): The atomic radii are computed from the atomic polarizabilities given by Nagle (1990).
7. The atomic electronegativity E-scheme (Ivanciuc, 2000a): The atomic electronegativities are computed from the atomic polarizabilities given by Nagle (1990).

Various distance indices can be computed using the Barysz (weighted) vertex-distance matrix.

## 4.11   THE COMPLEMENT OF THE BARYSZ VERTEX-DISTANCE MATRIX

The *complement* of the Barysz vertex-distance matrix, denoted by $^{cBv}\mathbf{D}$, is defined by means of the $^{Bv}\mathbf{D}$ matrix:

$$\left[ ^{cBv}\mathbf{D} \right]_{ij} = \begin{cases} V - \left[ ^{Bv}\mathbf{D} \right]_{ij} & \text{if } i \neq j \\ w_{ii} & \text{if } i = j \text{ and if the vertex } i \text{ is weighted} \\ 0 & \text{otherwise} \end{cases} \qquad (4.21)$$

The $^{cBv}\mathbf{D}$ matrix of $G_4$ is given as follows:

$$^{cBv}\mathbf{D}(G_4) = \begin{bmatrix} 0 & 6 & 5 & 4.143 & 3 & 4 & 3 \\ 6 & 0 & 6 & 5.143 & 4 & 5 & 4 \\ 5 & 6 & 0 & 6.143 & 5 & 6 & 5 \\ 4.143 & 5.143 & 6.143 & 0.143 & 6.143 & 5.143 & 4.143 \\ 3 & 4 & 5 & 6.143 & 0 & 6 & 5 \\ 4 & 5 & 6 & 5.143 & 6 & 0 & 6 \\ 3 & 4 & 5 & 4.143 & 5 & 6 & 0 \end{bmatrix}$$

The Wiener-like index based on this matrix was utilized by Ivanciuc in the QSAR modeling (Ivanciuc, 2000b).

## 4.12   THE RECIPROCAL BARYSZ VERTEX-DISTANCE MATRIX

The *reciprocal Barysz vertex-distance matrix*, denoted by $^{Bv}\mathbf{D}^{-1}$, of a vertex- and edge-weighted molecular graph $G$ with $V$ vertices is the square symmetric $V \times V$ matrix defined as

$$
\left[ ^{Bv}\mathbf{D}^{-1} \right]_{ij} = \begin{cases} 1/w_{ij} & \text{if } i \neq j \\ w_{ii} & \text{if } i = j \text{ and if the vertex } i \text{ is weighted} \\ 0 & \text{otherwise} \end{cases} \tag{4.22}
$$

A similar definition of the $^{Bv}\mathbf{D}^{-1}$ matrix of a vertex- and edge-weighted molecular graph was offered by Ivanciuc (2000c) and Ivanciuc et al. (1998, 1999).

The $^{Bv}\mathbf{D}^{-1}$ matrix of $G_4$ is shown below:

$$
^{Bv}\mathbf{D}^{-1}(G_4) = \begin{bmatrix} 0 & 1 & 0.50 & 0.35 & 0.25 & 0.33 & 0.25 \\ 1 & 0 & 1 & 0.54 & 0.33 & 0.50 & 0.33 \\ 0.50 & 1 & 0 & 1.17 & 0.50 & 1 & 0.50 \\ 0.35 & 0.54 & 1.17 & 0.143 & 1.17 & 0.54 & 0.35 \\ 0.25 & 0.33 & 0.50 & 1.17 & 0 & 1 & 0.50 \\ 0.33 & 0.50 & 1 & 0.54 & 1 & 0 & 1 \\ 0.25 & 0.33 & 0.50 & 0.35 & 0.50 & 1 & 0 \end{bmatrix}
$$

The sum of the entries in the upper matrix-triangle (or the lower matrix-triangle) gives the Wiener index of $G_4$. The molecular descriptors based on the reciprocal Barysz vertex-distance matrix have been used in QSAR modeling by Ivanciuc (2000b).

## 4.13   THE RECIPROCAL OF THE COMPLEMENT OF THE BARYSZ VERTEX-DISTANCE MATRIX

The *reciprocal* of the complement of the Barysz vertex-distance matrix, denoted by $1/^{cBv}\mathbf{D}$, of a vertex- and edge-weighted molecular graph $G$ with $V$ vertices, is the square $V \times V$ symmetric matrix and is defined as

$$
\left[ ^{cBv}\mathbf{D}^{-1} \right]_{ij} = \begin{cases} 1/[^{cBv}\mathbf{D}]_{ij} & \text{if } i \neq j \\ w_{ii} & \text{if } i = j \text{ and if the vertex } i \text{ is weighted} \\ 0 & \text{otherwise} \end{cases} \tag{4.23}
$$

The *reciprocal* of the complement of the Barysz vertex-distance matrix $^{cBv}\mathbf{D}^{-1}$ of $G_4$ follows:

$$
^{cBv}\mathbf{D}^{-1}(G_4) = \begin{bmatrix}
0 & 0.167 & 0.200 & 0.241 & 0.333 & 0.250 & 0.333 \\
0.167 & 0 & 0.167 & 0.194 & 0.250 & 0.200 & 0.250 \\
0.200 & 0.167 & 0 & 0.163 & 0.200 & 0.167 & 0.200 \\
0.241 & 0.194 & 0.163 & 0.143 & 0.163 & 0.194 & 0.241 \\
0.333 & 0.250 & 0.200 & 0.163 & 0 & 0.167 & 0.200 \\
0.250 & 0.200 & 0.167 & 0.194 & 0.167 & 0 & 0.167 \\
0.333 & 0.250 & 0.200 & 0.241 & 0.200 & 0.167 & 0
\end{bmatrix}
$$

The Wiener-like index based on the $^{cBv}\mathbf{D}^{-1}$ matrix has also been used by Ivanciuc (2000b, 2000c) in the QSAR modeling.

## 4.14   THE COMPLEMENTARY VERTEX-DISTANCE MATRIX

The *complementary vertex-distance matrix*, denoted by $^{cv}\mathbf{D}$, has been introduced by Ivanciuc (2000c) and discussed by Balaban et al. (2000) and Ivanciuc et al. (2000). It is a square symmetric $V \times V$ matrix defined as

$$
\left[ {}^{cv}\mathbf{D} \right]_{ij} = \begin{cases} l_{\min} + l_{\max} - [{}^{v}\mathbf{D}]_{ij} & \text{if } i \neq j \\ 0 & \text{otherwise} \end{cases}
\tag{4.24}
$$

where $l_{\min}$ and $l_{\max}$ are the minimum distance and the maximum distance in a graph. If a graph is simple, then $l_{\min} = 1$ and $l_{\max} =$ the graph diameter $D$ and the complementary vertex-distance matrix (4.24) becomes

$$
\left[ {}^{cv}\mathbf{D} \right]_{ij} = \begin{cases} 1 + D - [{}^{v}\mathbf{D}]_{ij} & \text{if } i \neq j \\ 0 & \text{otherwise} \end{cases}
\tag{4.25}
$$

The diameter $D$ of a graph $G$ is the longest geodesic distance between any two vertices $i$ and $j$ in $G$, i.e., the largest $[{}^{v}\mathbf{D}]_{ij}$ value in the vertex-distance matrix (Harary, 1971).

The elements of the complementary vertex-distance matrix differ from the elements of the reverse-Wiener matrix only for unity (see Section 5.5).

The complementary vertex-distance matrices of $T_2$ (see Figure 2.19) and $G_1$ (see structure $A$ in Figure 2.1) are as follows:

$$
{}^{cv}\mathbf{D}(T_2) = \begin{bmatrix}
0 & 5 & 4 & 3 & 2 & 1 & 3 & 4 \\
5 & 0 & 5 & 4 & 3 & 2 & 4 & 5 \\
4 & 5 & 0 & 5 & 4 & 3 & 5 & 4 \\
3 & 4 & 5 & 0 & 5 & 4 & 4 & 3 \\
2 & 3 & 4 & 5 & 0 & 5 & 3 & 2 \\
1 & 2 & 3 & 4 & 5 & 0 & 2 & 1 \\
3 & 4 & 5 & 4 & 3 & 2 & 0 & 3 \\
4 & 5 & 4 & 3 & 2 & 1 & 3 & 0
\end{bmatrix}
$$

$$
{}^{cv}\mathbf{D}(G_1) = \begin{bmatrix}
0 & 4 & 3 & 2 & 1 & 2 & 1 \\
4 & 0 & 4 & 3 & 2 & 3 & 2 \\
3 & 4 & 0 & 4 & 3 & 4 & 3 \\
2 & 3 & 4 & 0 & 4 & 3 & 2 \\
1 & 2 & 3 & 4 & 0 & 4 & 3 \\
2 & 3 & 4 & 3 & 4 & 0 & 4 \\
1 & 2 & 3 & 2 & 3 & 4 & 0
\end{bmatrix}
$$

The complementary vertex-distance matrices are used to generate the Wiener-like molecular descriptors that have been successfully tested in QSPR modeling (Ivanciuc et al., 2000). The complementary vertex-distance matrices of vertex- and edge-weighted graphs have also been introduced and used in QSPR (Ivanciuc, 2000c).

## 4.15  THE RECIPROCAL OF THE COMPLEMENTARY VERTEX-DISTANCE MATRIX

The reciprocal of the complementary vertex-distance matrix, denoted by ${}^{cv}\mathbf{D}^{-1}$, is simply given by

$$
{}^{cv}\mathbf{D}^{-1} = 1/{}^{cv}\mathbf{D} \tag{4.26}
$$

The reciprocal complementary vertex-distance matrices of $T_2$ (see Figure 2.19) and $G_1$ (see structure $A$ in Figure 2.1) are given below:

$$
{}^{cv}\mathbf{D}^{-1}(T_2) = \begin{bmatrix}
0 & 1/5 & 1/4 & 1/3 & 1/2 & 1 & 1/3 & 1/4 \\
1/5 & 0 & 1/5 & 1/4 & 1/3 & 1/2 & 1/4 & 1/5 \\
1/4 & 1/5 & 0 & 1/5 & 1/4 & 1/3 & 1/5 & 1/4 \\
1/3 & 1/4 & 1/5 & 0 & 1/5 & 1/4 & 1/4 & 1/3 \\
1/2 & 1/3 & 1/4 & 1/5 & 0 & 1/5 & 1/3 & 1/2 \\
1 & 1/2 & 1/3 & 1/4 & 1/5 & 0 & 1/2 & 1 \\
1/3 & 1/4 & 1/5 & 1/4 & 1/3 & 1/2 & 0 & 1/3 \\
1/4 & 1/5 & 1/4 & 1/3 & 1/2 & 1 & 1/3 & 0
\end{bmatrix}
$$

$$
{}^{cv}\mathbf{D}^{-1}(G_1) = \begin{bmatrix}
0 & 1/4 & 1/3 & 1/2 & 1 & 1/2 & 1 \\
1/4 & 0 & 1/4 & 1/3 & 1/2 & 1/3 & 1/2 \\
1/3 & 1/4 & 0 & 1/4 & 1/3 & 1/4 & 1/3 \\
1/2 & 1/3 & 1/4 & 0 & 1/4 & 1/3 & 1/2 \\
1 & 1/2 & 1/3 & 1/4 & 0 & 1/4 & 1/3 \\
1/2 & 1/3 & 1/4 & 1/3 & 1/4 & 0 & 1/4 \\
1 & 1/2 & 1/3 & 1/2 & 1/3 & 1/4 & 0
\end{bmatrix}
$$

Ivanciuc (2000c) extended the concept of reciprocal of the complementary vertex-distance matrix to the vertex- and edge-weighted graphs and used the derived Wiener-like indices in QSPR modeling.

## 4.16   MATRIX OF DOMINANT DISTANCES IN A GRAPH

Randić (2013) introduced the *matrix of dominant distances* in a graph and denoted it by $\mathbf{D}_{MAX}$. This matrix is constructed from the vertex-distance matrix ${}^v\mathbf{D}$ of a graph by selecting for each row and each column only the largest distances. The matrix of dominant distances in a graph can be considered as opposite to the vertex-adjacency matrix ${}^v\mathbf{A}$, which can be constructed from the distance matrix by selecting for each row and each column only the distances of the unity length. Below we give $\mathbf{D}_{MAX}$ matrices for $T_2$ (see Figure 2.19) and $G_1$ (see structure $A$ in Figure 2.1).

$$
\mathbf{D}_{max}(T_2) = \begin{bmatrix}
0 & 0 & 0 & 3 & 4 & 5 & 3 & 0 \\
0 & 0 & 0 & 0 & 0 & 4 & 0 & 0 \\
0 & 0 & 0 & 0 & 0 & 3 & 0 & 0 \\
3 & 0 & 0 & 0 & 0 & 0 & 0 & 3 \\
4 & 0 & 0 & 0 & 0 & 0 & 0 & 4 \\
5 & 4 & 3 & 0 & 0 & 0 & 4 & 5 \\
3 & 0 & 0 & 0 & 0 & 4 & 0 & 3 \\
0 & 0 & 0 & 3 & 4 & 5 & 3 & 0
\end{bmatrix}
$$

$$
\mathbf{D}_{max}(G_1) = \begin{bmatrix}
0 & 0 & 2 & 3 & 4 & 3 & 4 \\
0 & 0 & 0 & 0 & 3 & 0 & 3 \\
2 & 0 & 0 & 0 & 2 & 0 & 2 \\
3 & 0 & 0 & 0 & 0 & 0 & 3 \\
4 & 3 & 2 & 0 & 0 & 0 & 2 \\
3 & 0 & 0 & 0 & 0 & 0 & 0 \\
4 & 3 & 2 & 3 & 2 & 0 & 0
\end{bmatrix}
$$

The procedure to generate the $\mathbf{D}_{MAX}$ matrix is general enough that it may be applied to other distance-related matrices.

Randić et al. (2013a) explored the use of the $\mathbf{D}_{MAX}$ matrix for discrimination between graphs.

## 4.17   THE DETOUR MATRIX

The *detour matrix* (or the *maximum path matrix*) of a vertex-labeled connected graph $G$, denoted by $\mathbf{DM}$, is a real symmetric $V \times V$ matrix whose elements are defined as (Harary, 1971; Buckley and Harary, 1990; Amić and Trinajstić, 1995, Trinajstić et al., 1997a; Nikolić et al., 1999; Todeschini and Consonni, 2000, 2009)

$$\left[\mathbf{DM}\right]_{ij} = \begin{cases} L(i,j) & \text{if } i \neq j \\ 0 & \text{otherwise} \end{cases} \qquad (4.27)$$

where $L(i,j)$ is the length of the *longest* distance, i.e., the *maximum* number of edges, between vertices $i$ and $j$. The longest distance in a graph is called the *elongation*, and its length is equal to the *detour distance*, hence the term *detour matrix*. It is also convenient to call the longest path connecting the vertices $i$ and $j$ the *detour-path*. The detour matrix of the vertex-labeled graph $G_1$ (see structure $A$ in Figure 2.1) is

$$\mathbf{DM}(G_1) = \begin{bmatrix} 0 & 1 & 2 & 5 & 4 & 5 & 6 \\ 1 & 0 & 1 & 4 & 3 & 4 & 5 \\ 2 & 1 & 0 & 3 & 2 & 3 & 4 \\ 5 & 4 & 3 & 0 & 3 & 2 & 3 \\ 4 & 3 & 2 & 3 & 0 & 3 & 4 \\ 5 & 4 & 3 & 2 & 3 & 0 & 1 \\ 6 & 5 & 4 & 3 & 4 & 1 & 0 \end{bmatrix}$$

Rücker and Rücker (1998) proposed a slightly different definition of the detour matrix. These authors pointed out that it was never well explained by either Harary (1971) or others (e.g., Amić and Trinajstić, 1995) why zeros should appear as diagonal elements of the detour matrix. They therefore defined the diagonal elements of the detour matrix $[^{RR}\mathbf{DM}_{ii}]$ as the lengths of the shortest self-returning walks (*ssrw*), which visit all sites:

$$\left[^{RR}\mathbf{DM}\right]_{ij} = \begin{cases} L(i,j) & \text{if } i \neq j \\ ssrw(i,i) & \text{if } i = j \end{cases} \qquad (4.28)$$

Thus, the above detour matrix of the vertex-labeled graph $G_1$ (see structure $A$ in Figure 2.1) becomes

$$
{}^{RR}\mathbf{DM}(G_1) = \begin{bmatrix}
10 & 1 & 2 & 5 & 4 & 5 & 6 \\
1 & 10 & 1 & 4 & 3 & 4 & 5 \\
2 & 1 & 10 & 3 & 2 & 3 & 4 \\
5 & 4 & 3 & 10 & 3 & 2 & 3 \\
4 & 3 & 2 & 3 & 10 & 3 & 4 \\
5 & 4 & 3 & 2 & 3 & 10 & 1 \\
6 & 5 & 4 & 3 & 4 & 1 & 10
\end{bmatrix}
$$

where the diagonal elements represent the self-returning walks of length 10, which include visits to all the vertices in $G_1$.

While the standard distance matrix determines a graph uniquely, this is *not* the case with the detour matrix—there are nonisomorphic graphs with identical detour matrices. Two pairs of such graphs, taken from Randić et al. (1998), are shown in Figure 4.4.

The detour matrices belonging to the pairs of graphs in Figure 4.4 are given below:

$$
\mathbf{DM}(G_{19}) = \mathbf{DM}(G_{20}) = \begin{bmatrix}
0 & 3 & 4 & 3 & 4 \\
3 & 0 & 3 & 2 & 3 \\
4 & 3 & 0 & 3 & 4 \\
3 & 2 & 3 & 0 & 3 \\
4 & 3 & 4 & 3 & 0
\end{bmatrix}
$$

$$
\mathbf{DM}(G_{21}) = \mathbf{DM}(G_{22}) = \begin{bmatrix}
0 & 4 & 4 & 4 & 4 \\
4 & 0 & 4 & 4 & 4 \\
4 & 4 & 0 & 4 & 4 \\
4 & 4 & 4 & 0 & 4 \\
4 & 4 & 4 & 4 & 0
\end{bmatrix}
$$

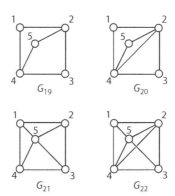

**FIGURE 4.4** Two pairs of nonisomorphic graphs, $G_{19}$-$G_{20}$ and $G_{21}$-$G_{22}$, with identical detour matrices.

Several methods are available for computing the detour matrix. We list three here:

1. The paper-and-pencil approach of path tracing on a graph $G$, which can be carried out by hand (Amić and Trinajstić, 1995) or with a computer (Lukovits and Razinger, 1997; Trinajstić et al., 1997b)
2. The utilization of the distance matrices of the spanning trees belonging to $G$ (Nikolić et al., 1996; Trinajstić et al., 1997a)
3. The symmetry-added computation (Rücker and Rücker, 1998)

We present here the second method by which the detour matrix of a polycyclic graph can be generated from the vertex-distance matrices of the corresponding spanning trees following the procedure consisting of the three steps outlined below (Trinajstić et al., 1997a; Nikolić et al., 1990, 1996):

1. Labeling the vertices of a polycyclic graph $G$ under consideration
2. Generation of labeled spanning trees from $G$ and construction of their vertex-distance matrices
3. Setting up the detour matrix of $G$ by matching the vertex-distance matrices of spanning trees, selecting only those elements that possess the largest numerical values, and placing them in the appropriate place in the detour matrix.

We illustrate this procedure by considering graph $G_1$. Its vertex-labels are given as structure $A$ in Figure 2.1. $G_1$ can have only four spanning trees—they are presented in Figure 2.22. Vertex-distance matrices belonging to spanning trees from Figure 2.22 are listed below:

$$
{}^{v}\mathbf{D}(ST_1) = \begin{bmatrix} 0 & 1 & 2 & 5 & 4 & 3 & 4 \\ 1 & 0 & 1 & 4 & 3 & 2 & 3 \\ 2 & 1 & 0 & 3 & 2 & 1 & 2 \\ 5 & 4 & 3 & 0 & 1 & 2 & 3 \\ 4 & 3 & 2 & 1 & 0 & 1 & 2 \\ 3 & 2 & 1 & 2 & 1 & 0 & 1 \\ 4 & 3 & 2 & 3 & 2 & 1 & 0 \end{bmatrix}
$$

$$
{}^{v}\mathbf{D}(ST_2) = \begin{bmatrix} 0 & 1 & 2 & 3 & 4 & 3 & 4 \\ 1 & 0 & 1 & 2 & 3 & 2 & 3 \\ 2 & 1 & 0 & 1 & 2 & 1 & 2 \\ 3 & 2 & 1 & 0 & 3 & 2 & 3 \\ 4 & 3 & 2 & 3 & 0 & 1 & 2 \\ 3 & 2 & 1 & 2 & 1 & 0 & 1 \\ 4 & 3 & 2 & 3 & 2 & 1 & 0 \end{bmatrix}
$$

$$
{}^{v}\mathbf{D}(ST_3) = \begin{bmatrix}
0 & 1 & 2 & 3 & 4 & 3 & 4 \\
1 & 0 & 1 & 2 & 3 & 2 & 3 \\
2 & 1 & 0 & 1 & 2 & 1 & 2 \\
3 & 2 & 1 & 0 & 1 & 2 & 3 \\
4 & 3 & 2 & 1 & 0 & 3 & 4 \\
3 & 2 & 1 & 2 & 3 & 0 & 1 \\
4 & 3 & 2 & 3 & 4 & 1 & 0
\end{bmatrix}
$$

$$
{}^{v}\mathbf{D}(ST_4) = \begin{bmatrix}
0 & 1 & 2 & 3 & 4 & 5 & 6 \\
1 & 0 & 1 & 2 & 3 & 4 & 5 \\
2 & 1 & 0 & 1 & 2 & 3 & 4 \\
3 & 2 & 1 & 0 & 1 & 2 & 3 \\
4 & 3 & 2 & 1 & 0 & 1 & 2 \\
5 & 4 & 3 & 2 & 1 & 0 & 1 \\
6 & 5 & 4 & 3 & 2 & 1 & 0
\end{bmatrix}
$$

Matching these four vertex-distance matrices and choosing the appropriate elements leads to the detour matrix of $G_1$ that was presented above. It should also be pointed out that the vertex-distance matrix and the detour matrix are identical for acyclic graphs.

The detour matrix was used to generate a Wiener-like index (Amić and Trinajstić, 1995), named the detour index (Lukovits, 1996), for simple and weighted graphs (Amić and Trinajstić, 1995; Nikolić et al., 1990, 1996, 1999; Lukovits, 1996; Trinajstić et al., 1997a, 1997b, 2001; Lukovits and Razinger, 1997). All kinds of distance indices can be generated from the detour matrix (Lukovits, 1996; Nikolić et al., 1999; Todeschini and Consonni, 2000, 2009; Lučić et al., 2001; Randić and Pompe, 2001; Trinajstić et al., 2001).

## 4.18   THE DETOUR-PATH MATRIX

The *detour-path matrix*, denoted by ${}^{p}\mathbf{DM}$, can similarly be defined as the vertex-distance-path matrix; that is, the matrix ${}^{p}\mathbf{DM}$ is a square symmetric $V \times V$ matrix whose off-diagonal elements $i, j$ count all paths of any length that are included within the longest path between vertex $i$ and vertex $j$ (Diudea, 1996a). Each element $i, j$ of the ${}^{p}\mathbf{DM}$ is computed from the corresponding detour matrix as follows:

$$
\left[{}^{p}\mathbf{DM}\right]_{ij} = \begin{cases}
\left( \dfrac{[\mathbf{DM}]_{ij} + 1}{2} \right) & \text{if } i \neq j \\
0 & \text{otherwise}
\end{cases}
\tag{4.29}
$$

The detour-path matrix of $G_1$ (see structure $A$ in Figure 2.1) is given below:

$$^{p}\mathbf{DM}(G_1) = \begin{bmatrix} 0 & 1 & 3 & 15 & 10 & 15 & 21 \\ 1 & 0 & 1 & 10 & 6 & 10 & 15 \\ 3 & 1 & 0 & 6 & 3 & 6 & 10 \\ 15 & 10 & 6 & 0 & 6 & 3 & 6 \\ 10 & 6 & 3 & 6 & 0 & 6 & 10 \\ 15 & 10 & 6 & 3 & 6 & 0 & 1 \\ 21 & 15 & 10 & 6 & 10 & 1 & 0 \end{bmatrix}$$

The hyper-detour index can be obtained from the detour-path matrix. For acyclic graphs, the detour-path matrix is equal to the distance-path matrix, and consequently, the hyper-detour index for acyclic graphs is equal to the hyper-distance-path index obtained from the distance-path matrix. The hyper-detour index has been used in the structure property modeling of lower acyclic and cyclic saturated hydrocarbons with up to eight carbon atoms (Trinajstić et al., 2001).

## 4.19   THE DETOUR-DELTA MATRIX

The entries in the *detour-delta matrix*, denoted by $^{\Delta}\mathbf{DM}$, are related to the elements of the corresponding detour matrix as the entries of the vertex-distance-delta matrix are to the elements of the vertex-distance matrix.

$$\left[^{\Delta}\mathbf{DM}\right]_{ij} = \begin{cases} \left( \dfrac{[\mathbf{DM}]_{ij}}{2} \right) & \text{if } i \neq j \\ 0 & \text{otherwise} \end{cases} \tag{4.30}$$

The detour-delta matrix enumerates the number of all *longest* paths larger than unity between vertices $i$ and $j$ in a graph. The $^{\Delta}\mathbf{DM}$ matrix of $G_1$ (see structure A in Figure 2.1) is given below:

$$^{\Delta}\mathbf{DM}(G_1) = \begin{bmatrix} 0 & 0 & 1 & 10 & 6 & 10 & 15 \\ 0 & 0 & 0 & 6 & 3 & 6 & 10 \\ 1 & 0 & 0 & 3 & 1 & 3 & 6 \\ 10 & 6 & 3 & 0 & 3 & 2 & 3 \\ 6 & 3 & 1 & 3 & 0 & 3 & 6 \\ 10 & 6 & 3 & 2 & 3 & 0 & 0 \\ 15 & 10 & 6 & 3 & 6 & 0 & 0 \end{bmatrix}$$

The following relationship is between the elements of the three matrices based on the longest graph-distances $\mathbf{DM}$, $^{p}\mathbf{DM}$, and $^{\Delta}\mathbf{DM}$:

$$[^{p}\mathbf{DM}]_{ij} = [\mathbf{DM}]_{ij} + [^{\Delta}\mathbf{DM}]_{ij} \tag{4.31}$$

## 4.20   THE EDGE-WEIGHTED DETOUR MATRIX

The edge-weighted detour matrix, denoted $^{ew}\mathbf{DM}$, is a square symmetric $V \times V$ matrix defined as (Nikolić et al., 1996)

$$\left[ ^{ew}\mathbf{DM} \right]_{ij} = \begin{cases} w(dp_{ij}) & \text{if } i \neq j \\ 0 & \text{otherwise} \end{cases} \tag{4.32}$$

where $w(dp_{ij})$ is the maximum sum of edge-weights along the detour-path between the vertices $i$ and $j$ in an edge-weighted graph $G_{ew}$, which may not be the *longest* possible detour-path between these two vertices in $G_{ew}$. Consequently, in the case of the detour matrix for edge-weighted graphs, the entry $[^{ew}\mathbf{DM}]_{ij}$ is the maximum detour-path-weight between the vertices $i$ and $j$. The edge-weighted detour matrix of $G_{18}$ (see Figure 4.2) is shown below:

$$^{ew}\mathbf{DM}(G_{18}) = \begin{bmatrix} 0 & 2 & 3 & 7 & 6 & 7 & 9 \\ 2 & 0 & 1 & 5 & 4 & 5 & 7 \\ 3 & 1 & 0 & 4 & 3 & 4 & 6 \\ 7 & 5 & 4 & 0 & 3 & 3 & 5 \\ 6 & 4 & 3 & 3 & 0 & 4 & 6 \\ 7 & 5 & 4 & 3 & 4 & 0 & 2 \\ 9 & 7 & 6 & 5 & 6 & 2 & 0 \end{bmatrix}$$

By analogy with the construction of the detour matrix of polycyclic graphs from the distance matrices of the corresponding spanning trees, the same approach can be applied to composing the edge-weighted detour matrix of weighted polycyclic graphs from the corresponding edge-weighted spanning trees (Nikolić et al., 1990). This approach is illustrated for $G_{18}$. The four edge-weighted spanning trees of $G_{18}$, denoted by $^{ew}ST_n$ ($n = 1, 2, 3, 4$), are presented in Figure 4.5.

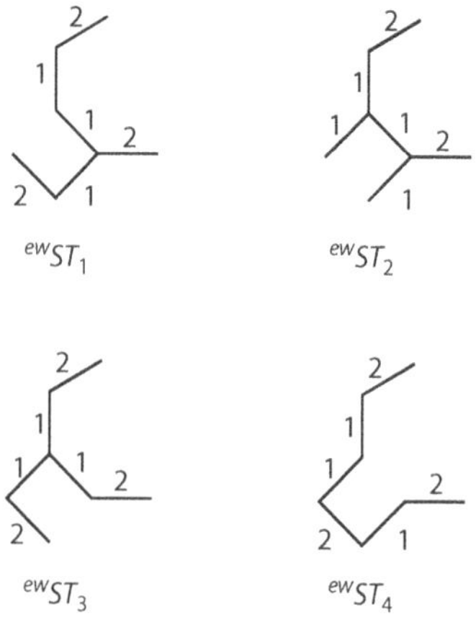

**FIGURE 4.5**   Edge-weighted spanning trees of $G_{18}$.

The edge-weighted vertex-distance matrices of four $^{ew}\mathrm{ST}_n$ ($n = 1, 2, 3, 4$) are given below:

$$
^{ew}\mathbf{D}(ST_1) =
\begin{bmatrix}
0 & 2 & 3 & 7 & 5 & 4 & 6 \\
2 & 0 & 1 & 5 & 3 & 2 & 4 \\
3 & 1 & 0 & 4 & 2 & 1 & 3 \\
7 & 5 & 4 & 0 & 2 & 3 & 5 \\
5 & 3 & 2 & 2 & 0 & 1 & 3 \\
4 & 2 & 1 & 3 & 1 & 0 & 2 \\
6 & 4 & 3 & 5 & 3 & 2 & 0
\end{bmatrix}
$$

$$
^{ew}\mathbf{D}(ST_2) =
\begin{bmatrix}
0 & 2 & 3 & 4 & 5 & 4 & 6 \\
2 & 0 & 1 & 2 & 3 & 2 & 4 \\
3 & 1 & 0 & 1 & 2 & 1 & 3 \\
4 & 2 & 1 & 0 & 3 & 2 & 4 \\
5 & 3 & 2 & 3 & 0 & 1 & 3 \\
4 & 2 & 1 & 2 & 1 & 0 & 2 \\
6 & 4 & 3 & 4 & 3 & 2 & 0
\end{bmatrix}
$$

$$
^{ew}\mathbf{D}(ST_3) =
\begin{bmatrix}
0 & 2 & 3 & 4 & 6 & 4 & 6 \\
2 & 0 & 1 & 2 & 4 & 2 & 4 \\
3 & 1 & 0 & 1 & 3 & 1 & 3 \\
4 & 2 & 1 & 0 & 2 & 2 & 4 \\
6 & 4 & 3 & 2 & 0 & 4 & 6 \\
4 & 2 & 1 & 2 & 4 & 0 & 2 \\
6 & 4 & 3 & 4 & 6 & 2 & 0
\end{bmatrix}
$$

$$
^{ew}\mathbf{D}(ST_4) =
\begin{bmatrix}
0 & 2 & 3 & 4 & 6 & 7 & 9 \\
2 & 0 & 1 & 2 & 4 & 5 & 7 \\
3 & 1 & 0 & 1 & 3 & 4 & 6 \\
4 & 2 & 1 & 0 & 2 & 3 & 5 \\
6 & 4 & 3 & 2 & 0 & 1 & 3 \\
7 & 5 & 4 & 3 & 1 & 0 & 2 \\
9 & 7 & 6 & 5 & 3 & 2 & 0
\end{bmatrix}
$$

Matching entries of these four matrices and selecting always the largest entry gives the edge-weighted detour matrix of $G_{18}$ shown above. Summation of the elements in the upper (or lower) matrix-triangle gives the weighted-detour index.

## 4.21   THE MAXIMUM-MINIMUM PATH MATRIX

The maximum-minimum path matrix, denoted by $DM - {}^{v}D$, has been introduced by Ivanciuc and Balaban (1994). This matrix is defined as

$$
\left[ DM - {}^{v}D \right]_{ij} = \begin{cases} L(i,j) & \text{if } i < j \\ 0 & \text{if } i = j \\ l(i,j) & \text{if } i > j \end{cases} \tag{4.33}
$$

where $L(i, j)$ and $l(i, j)$ are the elongation and the geodesic distance, as above, respectively. Thus, the upper triangle of the maximum-minimum path matrix contains the elements of the maximum-path matrix (detour matrix), and the lower triangle the elements of the minimum-path matrix (vertex-distance matrix). An example of the $DM - {}^{v}D$ matrix is given below for $G_1$ (see structure $A$ in Figure 2.1):

$$
DM - {}^{v}D(G_1) = \begin{bmatrix} 0 & 1 & 2 & 5 & 4 & 5 & 6 \\ 1 & 0 & 1 & 4 & 3 & 4 & 5 \\ 2 & 1 & 0 & 3 & 2 & 3 & 4 \\ 3 & 2 & 1 & 0 & 3 & 2 & 3 \\ 4 & 3 & 2 & 1 & 0 & 3 & 4 \\ 3 & 2 & 1 & 2 & 1 & 0 & 1 \\ 4 & 3 & 2 & 3 & 2 & 1 & 0 \end{bmatrix}
$$

The transpose of the matrix $DM - {}^{v}D$ is the minimum-maximum path matrix ${}^{v}D - DM$. The ${}^{v}D - DM$ matrix for $G_1$ is as follows:

$$
{}^{v}D - DM(G_1) = \begin{bmatrix} 0 & 1 & 2 & 3 & 4 & 3 & 4 \\ 1 & 0 & 1 & 2 & 3 & 2 & 3 \\ 2 & 1 & 0 & 1 & 2 & 1 & 2 \\ 5 & 4 & 3 & 0 & 1 & 2 & 3 \\ 4 & 3 & 2 & 3 & 0 & 1 & 2 \\ 5 & 4 & 3 & 2 & 3 & 0 & 1 \\ 6 & 5 & 4 & 3 & 4 & 1 & 0 \end{bmatrix}
$$

## 4.22   THE DETOUR-COMPLEMENT MATRIX

In parallel to the distance-complement matrix, the *detour-complement matrix*, denoted by ${}^{c}DM$, can be defined as

$$
\left[ {}^{c}DM \right]_{ij} = \begin{cases} V - [DM]_{ij} & \text{if } i \neq j \\ 0 & \text{otherwise} \end{cases} \tag{4.34}
$$

where $V$ is the number of vertices in a graph $G$.

The detour-complement matrix of the vertex-labeled graph $G_1$ (structure $A$ in Figure 2.1) is

$$
{}^c\mathbf{DM}(G_1) =
\begin{bmatrix}
0 & 6 & 5 & 2 & 3 & 2 & 1 \\
6 & 0 & 6 & 3 & 4 & 3 & 2 \\
5 & 6 & 0 & 4 & 5 & 4 & 3 \\
2 & 3 & 4 & 0 & 4 & 5 & 4 \\
3 & 4 & 5 & 4 & 0 & 4 & 3 \\
2 & 3 & 4 & 5 & 4 & 0 & 6 \\
1 & 2 & 3 & 4 & 3 & 6 & 0
\end{bmatrix}
$$

The detour-complement matrix can be used to generate the detour-complement index.

## 4.23  THE VERTEX-DISTANCE MATRIX AND THE DETOUR MATRIX OF COMPLETE GRAPHS AND COMPLETE BIPARTITE GRAPHS

A simple graph in which each pair of vertices is adjacent is called a *complete graph*. The complete graph on $V$ vertices is denoted by $K_V$. The number of edges in $K_V$ is $V(V-1)/2$. It should also be noted that the degree of each vertex in $K_V$ is $V-1$. As an example, we give in Figure 4.6 the complete graph on five vertices, denoted by $K_5$.

$K_5$ is one of the famous graphs of graph theory (Biggs et al., 1977), but it has also been used in chemistry, for example, in the study of the rearrangement of tetragonal-pyramidal complexes (e.g., Randić et al., 1983). It is also called the *Kuratowski graph* after the well-known Polish graph-theorist Kazimierz Kuratowski (Warsaw, 1896–Warsaw, 1980), who used this graph and $K_{3,3}$ graph in his studies on the graph planarity (Kuratowski, 1930). He has shown that a graph is planar if and only if it has no subgraph homomorphic to $K_5$ and $K_{3,3}$.

The $K_{3,3}$ graph is shown in Figure 4.7. This graph belongs to a family of *complete bipartite graphs* denoted by $K_{V°,V*}$, in which the set of $V$ vertices is partitioned into two subsets $V°$ and $V*$ such that the vertices in one subset are only joined with vertices in the other subset. It should also be noted that each of the vertices in, for example, the $V°$ set, is joined by a single edge to every vertex in the $V*$ set. The $K_{3,3}$ graph is also a well-known graph, even to nonmathematicians, since it often appears in puzzles (Dudeney, 1917; Lloyd, 1928). It is also known as the *utilities*

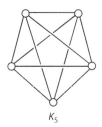

$K_5$

**FIGURE 4.6**   The $K_5$ graph.

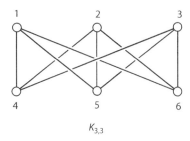

$K_{3,3}$

**FIGURE 4.7**   The labeled $K_{3,3}$ graph.

*graph* (Biggs et al., 1977). The utility graph is grounded in a puzzle discussed by Dudeney in his book *Amusements in Mathematics* (Dudeney, 1917), in which he collected a number of mathematically based puzzles. Dudeney presented a puzzle that can be visualized by the utility graph under the title *Water, Gas, and Electricity*, hence the name of the graph. The puzzle is to lay water, gas, and electricity pipes from three sources to each of three houses without any pipe crossing another.

The $K_5$ and $K_{3,3}$ graphs were also used in the discussion of the topological chirality of proteins (e.g., Liang and Mislow, 1994). In Euclidean 3-D space, a graph is topologically chiral if it cannot be converted to its mirror image by continuous deformation avoiding edge intersections. Nonplanarity is a necessary condition for topological chirality because a planar graph is achiral in 3-D space.

The vertex-distance matrix of a complete graph $K_V$ is defined as

$$\left[{}^{v}\mathbf{D}(K_V)\right]_{ij} = \begin{cases} V-d & \text{if } i \neq j \\ 0 & \text{otherwise} \end{cases} \tag{4.35}$$

Since $V - d$ is always equal to 1 in $K_V$ graphs, the corresponding vertex-distance matrices have all elements identical and equal to unity. Consequently, the vertex-distance matrices of $K_V$ graphs are identical to the vertex-adjacency matrices of these graphs. This is exemplified in the case of $K_5$.

$${}^{v}\mathbf{D}(K_5) = {}^{v}\mathbf{A}(K_5) = \begin{bmatrix} 0 & 1 & 1 & 1 & 1 \\ 1 & 0 & 1 & 1 & 1 \\ 1 & 1 & 0 & 1 & 1 \\ 1 & 1 & 1 & 0 & 1 \\ 1 & 1 & 1 & 1 & 0 \end{bmatrix}$$

The related Wiener index of complete graphs is then simply equal to the number of edges $E$ (Bonchev and Trinajstić, 1977).

The vertex-distance matrix of a complete bipartite graph $K_{V^{\circ},V^{*}}$ is given by

$$\left[ {}^{\gamma}\mathbf{D}(K_{V^\circ,V^*}) \right]_{ij} = \begin{cases} 1 & \text{if } i \text{ and } j \text{ are connected} \\ 2 & \text{if } i \text{ and } j \text{ are not connected} \\ 0 & \text{otherwise} \end{cases} \qquad (4.36)$$

As an example, we show below the ${}^{\gamma}\mathbf{D}$ matrix for $K_{3,3}$:

$${}^{\gamma}\mathbf{D}(K_{3,3}) = \begin{bmatrix} 0 & 2 & 2 & 1 & 1 & 1 \\ 2 & 0 & 2 & 1 & 1 & 1 \\ 2 & 2 & 0 & 1 & 1 & 1 \\ 1 & 1 & 1 & 0 & 2 & 2 \\ 1 & 1 & 1 & 2 & 0 & 2 \\ 1 & 1 & 1 & 2 & 2 & 0 \end{bmatrix}$$

The detour matrices of the complete graphs $K_V$ have off-diagonal elements equal to their vertex-degrees:

$$\left[ \mathbf{DM}(K_V) \right]_{ij} = \begin{cases} d & \text{if } i \neq j \\ 0 & \text{otherwise} \end{cases} \qquad (4.37)$$

The detour matrix of the $K_5$ graph is as follows:

$$\mathbf{DM}(K_5) = \begin{bmatrix} 0 & 4 & 4 & 4 & 4 \\ 4 & 0 & 4 & 4 & 4 \\ 4 & 4 & 0 & 4 & 4 \\ 4 & 4 & 4 & 0 & 4 \\ 4 & 4 & 4 & 4 & 0 \end{bmatrix}$$

The related detour index of complete graphs is the product of the number of edges and the vertex-degrees, i.e., $E \times d$.

In the case of the complete bipartite graphs $K_{V^\circ,V^*}$, the off-diagonal elements of the detour matrix are all equal to $V - 1$ or $V - 2$, depending on whether vertices are connected or not:

$$\left[ \mathbf{DM}(K_{V^\circ,V^*}) \right]_{ij} = \begin{cases} V - 1 & \text{if } i \text{ and } j \text{ are connected} \\ V - 2 & \text{if } i \text{ and } j \text{ are not connected} \\ 0 & \text{otherwise} \end{cases} \qquad (4.38)$$

The detour matrix of $K_{3,3}$ is as follows:

$$\mathbf{DM}(K_{3,3}) = \begin{bmatrix} 0 & 4 & 4 & 5 & 5 & 5 \\ 4 & 0 & 4 & 5 & 5 & 5 \\ 4 & 4 & 0 & 5 & 5 & 5 \\ 5 & 5 & 5 & 0 & 4 & 4 \\ 5 & 5 & 5 & 4 & 0 & 4 \\ 5 & 5 & 5 & 4 & 4 & 0 \end{bmatrix}$$

## 4.24  THE VERTEX-HARARY MATRIX

The *reciprocal vertex-distance matrix*, denoted by $^v\mathbf{D}^r$, is called the *vertex-Harary matrix* (Plavšić et al., 1993) in honor of the late Professor Frank Harary (New York, 1921–Las Cruces, New Mexico, 2004), the grandmaster of both graph theory and chemical graph theory. It can simply be obtained by replacing off-diagonal elements of the vertex-distance matrix $^v\mathbf{D}$ by their reciprocals (Balaban et al., 1992; Mihalić and Trinajstić, 1992; Plavšić et al., 1993; Ivanciuc et al., 1993; Todeschini and Consonni, 2000, 2009; Lučić et al., 2012):

$$\left[ ^v\mathbf{D}^r \right]_{ij} = \begin{cases} 1/[^v\mathbf{D}]_{ij} & \text{if } i \neq j \\ 0 & \text{otherwise} \end{cases} \tag{4.39}$$

As an example, we give below the vertex-Harary matrix for the vertex-labeled graph $G_1$ (see structure $A$ in Figure 2.1):

$$^v\mathbf{D}^r(G_1) = \begin{bmatrix} 0 & 1 & 1/2 & 1/3 & 1/4 & 1/3 & 1/4 \\ 1 & 0 & 1 & 1/2 & 1/3 & 1/2 & 1/3 \\ 1/2 & 1 & 0 & 1 & 1/2 & 1 & 1/2 \\ 1/3 & 1/2 & 1 & 0 & 1 & 1/2 & 1/3 \\ 1/4 & 1/3 & 1/2 & 1 & 0 & 1 & 1/2 \\ 1/3 & 1/2 & 1 & 1/2 & 1 & 0 & 1 \\ 1/4 & 1/3 & 1/2 & 1/3 & 1/2 & 1 & 0 \end{bmatrix}$$

A variant of the vertex-Harary matrix of a given graph is its squared version, denoted by $(^v\mathbf{D}^r)^2$, which is derived from the related vertex-distance matrix $^v\mathbf{D}$ by replacing its elements with the squares of their reciprocals (Plavšić et al., 1993; Consonni and Todeschini, 2012):

$$\left[ (^v\mathbf{D}^r)^2 \right]_{ij} = \begin{cases} 1/[^v\mathbf{D}]_{ij}^{\,2} & \text{if } i \neq j \\ 0 & \text{otherwise} \end{cases} \tag{4.40}$$

An example of this matrix for $G_1$ (see structure $A$ in Figure 2.1) is presented below:

$$(^v\mathbf{D}^r)^2(G_1) = \begin{bmatrix} 0 & 1 & 1/4 & 1/9 & 1/16 & 1/9 & 1/16 \\ 1 & 0 & 1 & 1/4 & 1/9 & 1/4 & 1/9 \\ 1/4 & 1 & 0 & 1 & 1/4 & 1 & 1/4 \\ 1/9 & 1/4 & 1 & 0 & 1 & 1/4 & 1/9 \\ 1/16 & 1/9 & 1/4 & 1 & 0 & 1 & 1/4 \\ 1/9 & 1/4 & 1 & 1/4 & 1 & 0 & 1 \\ 1/16 & 1/9 & 1/4 & 1/9 & 1/4 & 1 & 0 \end{bmatrix}$$

The motivation for introduction of the Harary matrix was pragmatic. The aim was to make a distance index differing from the Wiener index in the contributions of the distant sites. Namely, they should have much smaller contributions than the nearer sites, since in many circumstances the distant sites influence each other much less than nearer sites.

Soon after the introduction of the vertex-Harary index, the hyper-Harary index was proposed (Diudea, 1997). Then, the vertex-Harary index was extended to heterosystems (Ivanciuc et al., 1998), and finally, the modified vertex-Harary index was proposed and used (Lučić et al., 2002). The vertex-Harary index based on the matrix $(^v\mathbf{D}^r)^2$ was also successfully tested in several structure-property relationships (Mihalić and Trinajstić, 1992; Plavšić et al., 1993; Trinajstić et al., 2001). The vertex-Harary matrix can also be used to derive a variant of the Balaban index (Balaban, 1989). This novel index we call the Harary-Balaban index (Nikolić et al., 2001a).

## 4.25   THE EDGE-HARARY MATRIX

The *edge-Harary matrix* of a graph $G$, denoted by $^e\mathbf{D}^r$, is the vertex-Harary matrix of the corresponding line graph $L(G)$:

$$^e\mathbf{D}^r(G) = {}^v\mathbf{D}^r[L(G)] \tag{4.41}$$

We show below the edge-Harary matrix of $G_1$ (see structure $B$ in Figure 2.1), that is, the vertex-Harary matrix of $L(G_1)$ (see Figure 2.12).

$$^e\mathbf{D}^r(G_1) = \begin{bmatrix} 0 & 1 & 1/2 & 1/3 & 1/3 & 1/3 & 1/2 \\ 1 & 0 & 1 & 1/2 & 1/2 & 1/2 & 1 \\ 1/2 & 1 & 0 & 1 & 1/2 & 1/2 & 1 \\ 1/3 & 1/2 & 1 & 0 & 1 & 1/2 & 1/2 \\ 1/3 & 1/2 & 1/2 & 1 & 0 & 1 & 1 \\ 1/3 & 1/2 & 1/2 & 1/2 & 1 & 0 & 1 \\ 1/2 & 1 & 1 & 1/2 & 1 & 1 & 0 \end{bmatrix}$$

## 4.26   THE EDGE-WEIGHTED-HARARY MATRIX

A variant of the edge-Harary matrix, called the *edge-weighted-Harary matrix* and denoted by $^{ew}\mathbf{D}^r$, is a sparse square $V \times V$ matrix that is constructed in the following way. Any path in the graph can be broken into contributions of individual edges that make up that path. The same is true for reciprocal paths. It is convenient to call these edge-contributions the edge-weights. Then, the $i, j$-element of $^{ew}\mathbf{D}^r$ is equal to the sum of the weights of all individual $i$-$j$ edges making up paths of different lengths. This is illustrated for $T_2$ (see Figure 3.1) in Figure 4.8.

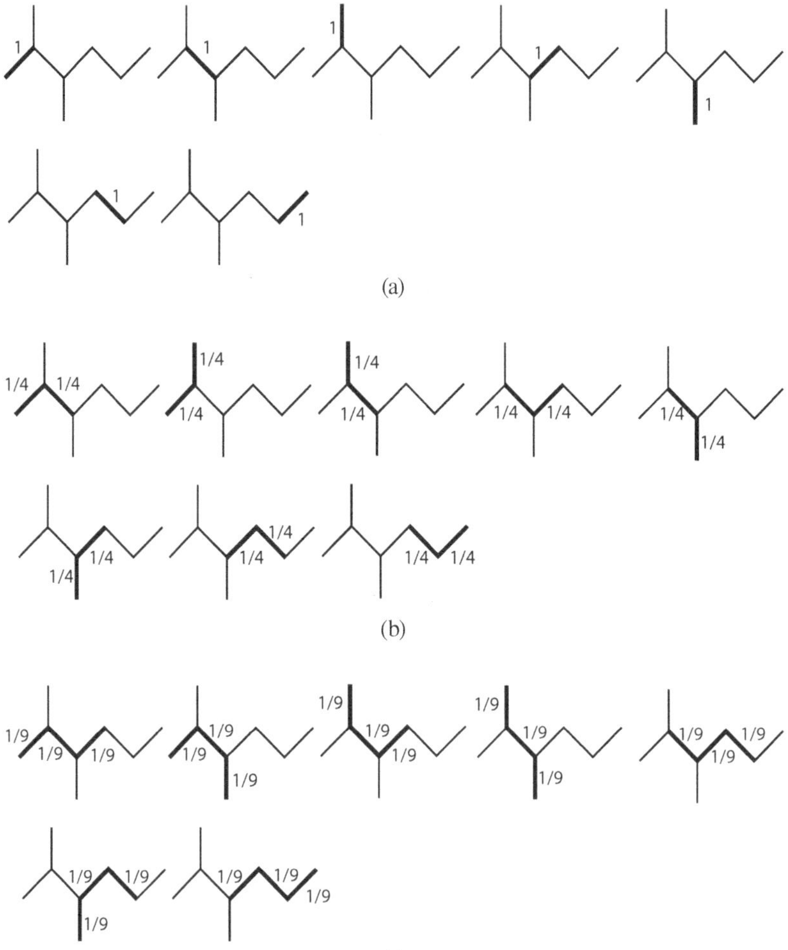

**FIGURE 4.8**   Weights of individual edges making up paths of different lengths that are used in the computation of the edge-weighted Harary matrix for $T_2$. (a) Weighted paths of length 1. (b) Weighted paths of length 2. (c) Weighted paths of length 3. (d) Weighted paths of length 4. (e) Weighted paths of length 5. *(continued)*

(d)

(e)

**FIGURE 4.8 (continued)** Weights of individual edges making up paths of different lengths that are used in the computation of the edge-weighted Harary matrix for $T_2$. (a) Weighted paths of length 1. (b) Weighted paths of length 2. (c) Weighted paths of length 3. (d) Weighted paths of length 4. (e) Weighted paths of length 5.

The edge-weighted-Harary matrix of $T_2$ is given as follows:

$$
{}^{ew}\mathbf{D}^r(T_2) = \begin{bmatrix}
0 & 1.825 & 0 & 0 & 0 & 0 & 0 & 0 \\
1.825 & 0 & 2.823 & 0 & 0 & 0 & 0 & 1.825 \\
0 & 2.823 & 0 & 2.636 & 0 & 0 & 1.896 & 0 \\
0 & 0 & 2.636 & 0 & 2.163 & 0 & 0 & 0 \\
0 & 0 & 0 & 2.163 & 0 & 1.566 & 0 & 0 \\
0 & 0 & 0 & 0 & 1.566 & 0 & 0 & 0 \\
0 & 0 & 1.896 & 0 & 0 & 0 & 0 & 0 \\
0 & 1.825 & 0 & 0 & 0 & 0 & 0 & 0
\end{bmatrix}
$$

The summation of the elements in the upper (or lower) half of the edge-weighted-Harary matrix gives the edge-Harary index. However, the edge-Harary index is identical to the vertex-Harary index for acyclic structures. In the case of the cycle-containing graphs, the edge-Harary index differs from the vertex-Harary index. To confirm this, we give in Figure 4.9 the weights of individual edges making up the paths in $G_1$ (see structure $A$ in Figure 2.1), and below the figure its edge-weighted-Harary matrix is given.

$$
{}^{ew}\mathbf{D}^r(G_1) = \begin{bmatrix}
0 & 1.768 & 0 & 0 & 0 & 0 & 0 \\
1.768 & 0 & 2.726 & 0 & 0 & 0 & 0 \\
0 & 2.726 & 0 & 2.735 & 0 & 3.068 & 0 \\
0 & 0 & 2.735 & 0 & 2.476 & 0 & 0 \\
0 & 0 & 0 & 2.476 & 0 & 2.663 & 0 \\
0 & 0 & 3.068 & 0 & 2.663 & 0 & 2.049 \\
0 & 0 & 0 & 0 & 0 & 2.049 & 0
\end{bmatrix}
$$

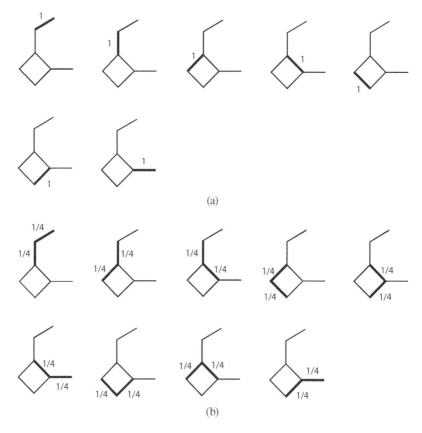

(a)

(b)

**FIGURE 4.9** Weights of edges making up paths of different lengths that are used for computing the edge-Harary matrix of $G_1$. (a) Weighted paths of length 1. (b) Weighted paths of length 2. (c) Weighted paths of length 3. (d) Weighted paths of length 4. (e) Weighted paths of length 5. (f) Weighted paths of length 6. *(continued)*

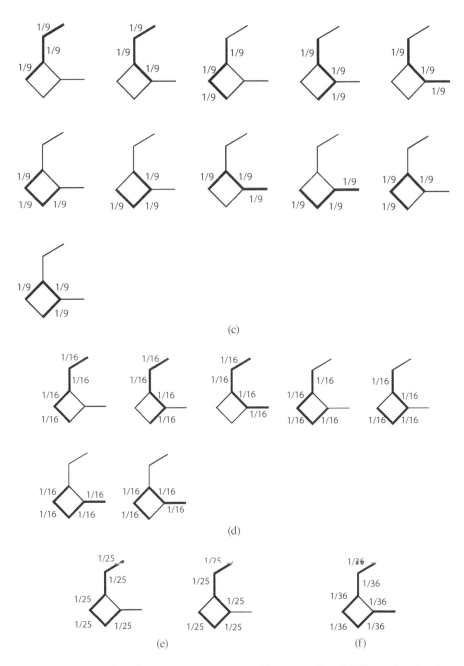

**FIGURE 4.9 (continued)** Weights of edges making up paths of different lengths that are used for computing the edge-Harary matrix of $G_1$. (a) Weighted paths of length 1. (b) Weighted paths of length 2. (c) Weighted paths of length 3. (d) Weighted paths of length 4. (e) Weighted paths of length 5. (f) Weighted paths of length 6.

The edge-weighted-Harary index for cycle-containing structures has not so far been tested in QSPR or QSAR modeling.

## 4.27  THE MODIFIED EDGE-WEIGHTED-HARARY MATRIX

The *modified edge-weighted-Harary matrix*, denoted by $^{mew}\mathbf{D}^r$, is obtained from the edge-weighted-Harary matrix by

$$
\left[ ^{mew}\mathbf{D}^r \right]_{ij} = \begin{cases} 1/[\,^{ew}\mathbf{D}^r]_{ij} & \text{if } i \neq j \\ 0 & \text{otherwise} \end{cases}
\tag{4.42}
$$

The modified edge-weighted-Harary matrices of $T_2$ (see Figure 3.1) and $G_1$ (see structure $A$ in Figure 2.1) are presented below:

$$
^{mew}\mathbf{D}^r(T_2) = \begin{bmatrix}
0 & 0.548 & 0 & 0 & 0 & 0 & 0 & 0 \\
0.548 & 0 & 0.354 & 0 & 0 & 0 & 0 & 0.548 \\
0 & 0.354 & 0 & 0.379 & 0 & 0 & 0.527 & 0 \\
0 & 0 & 0.379 & 0 & 0.462 & 0 & 0 & 0 \\
0 & 0 & 0 & 0.462 & 0 & 0.639 & 0 & 0 \\
0 & 0 & 0 & 0 & 0.639 & 0 & 0 & 0 \\
0 & 0 & 0.527 & 0 & 0 & 0 & 0 & 0 \\
0 & 0.548 & 0 & 0 & 0 & 0 & 0 & 0
\end{bmatrix}
$$

$$
^{mew}\mathbf{D}^r(G_1) = \begin{bmatrix}
0 & 0.566 & 0 & 0 & 0 & 0 & 0 \\
0.566 & 0 & 0.367 & 0 & 0 & 0 & 0 \\
0 & 0.367 & 0 & 0.366 & 0 & 0.326 & 0 \\
0 & 0 & 0.366 & 0 & 0.404 & 0 & 0 \\
0 & 0 & 0 & 0.404 & 0 & 0.376 & 0 \\
0 & 0 & 0.326 & 0 & 0.376 & 0 & 0.488 \\
0 & 0 & 0 & 0 & 0 & 0.488 & 0
\end{bmatrix}
$$

The modified edge-weighted-Harary matrix serves for computing the modified edge-weighted-Harary index. This index has been successfully tested in the structure-property modeling of alkanes (Lučić et al., 2002), but its use with polycyclic structures has not yet been investigated.

## 4.28  DISTANCE-DEGREE MATRICES

There are two kinds of the *distance-degree matrices*: one is based on the vertex-distance matrix and the vertex-degrees, and the other is based on the edge-distance matrix and edge-degrees.

The *vertex-distance-vertex-degree matrix* of a simple graph $G$ with $V$ vertices, denoted by $^vD^vd$ $(p, q, r)$, is a square $V \times V$ matrix defined as follows (Ivanciuc, 1999, 2000c):

$$\left[ ^vD^vd(p,q,r) \right]_{ij} = \begin{cases} l^p(ij)d^q(i)d^r(j) & \text{if } i \neq j \\ 0 & \text{otherwise} \end{cases}$$ (4.43)

where, as before, $l(ij)$ is the shortest graph-theoretical distance between vertices $i$ and $j$, and $d(i)$ and $d(j)$ are the degrees of vertices $i$ and $j$. The parameters $p$, $q$, and $r$ are natural numbers. The structure of a particular vertex-distance-vertex-degree matrix depends on the selected numerical values of these parameters. For example, if we choose the following values for the parameters, $p = 1$, $q = 0$, and $r = 0$, then the $^vD^vd$ matrix obtained will be identical to the vertex-distance matrix. If, on the other hand, we choose a slightly different set of parameter values, such as $p = -1$, $q = 0$, and $r = 0$, then the $^vD^vd$ matrix obtained is identical with the vertex-Harary matrix. As an example of computing the vertex-distance-vertex-degree matrix, we give below this matrix for $T_2$ (see Figure 2.19) with parameters $p = 1$, $q = 1$, and $r = 1$:

$$^vD^vd(1,1,1,T_2) = \begin{bmatrix} 0 & 3 & 6 & 6 & 8 & 5 & 3 & 2 \\ 3 & 0 & 9 & 12 & 18 & 12 & 6 & 3 \\ 6 & 9 & 0 & 6 & 12 & 9 & 3 & 6 \\ 6 & 12 & 6 & 0 & 4 & 4 & 4 & 6 \\ 8 & 18 & 12 & 4 & 0 & 2 & 6 & 8 \\ 5 & 12 & 9 & 4 & 2 & 0 & 4 & 5 \\ 3 & 6 & 3 & 4 & 6 & 4 & 0 & 3 \\ 2 & 3 & 6 & 6 & 8 & 5 & 3 & 0 \end{bmatrix}$$

From the definition of vertex-distance-vertex-degree matrix, it can be seen that nonsymmetric $^vD^vd$ matrices are obtained if $q \neq r$. As an example, consider $G_1$ (see structure A on Figure 2.1) with parameters $p = 1$, $q = 2$, and $r = 1$,

$$^vD^vd(1,2,1,G_1) = \begin{bmatrix} 0 & 2 & 6 & 6 & 8 & 9 & 4 \\ 4 & 0 & 12 & 16 & 24 & 24 & 12 \\ 18 & 18 & 0 & 18 & 36 & 27 & 18 \\ 12 & 16 & 12 & 0 & 8 & 24 & 12 \\ 16 & 24 & 24 & 8 & 0 & 12 & 8 \\ 27 & 36 & 27 & 36 & 18 & 0 & 9 \\ 4 & 6 & 6 & 6 & 4 & 3 & 0 \end{bmatrix}$$

The *edge-distance-edge-degree matrix* of a simple graph $G$ with $V$ vertices, denoted by ${}^e\mathbf{D}^e\mathbf{d}$ $(p, q, r)$, is the vertex-distance-vertex-degree matrix of the corresponding line graph $L(G)$. If we select the following values of parameters, $p = 1$, $q = 0$, and $r = 0$, the resulting ${}^e\mathbf{D}^e\mathbf{d}$ matrix of $G$ will be identical to the edge-distance matrix of $L(G)$. Selection of the parameters $p = -1$, $q = 0$, and $r = 0$ yields the ${}^e\mathbf{D}^e\mathbf{d}$ matrix of $G$ identical to the vertex-Harary matrix of $L(G)$. Below we give the ${}^e\mathbf{D}^e\mathbf{d}$ matrix of $G_1$ for the parameters selection $p = -1$, $q = 0$, and $r = 0$.

$$
{}^e\mathbf{D}^e\mathbf{d}(-1,0,0,G_1) =
\begin{bmatrix}
0 & 1 & 1/2 & 1/3 & 1/3 & 1/3 & 1/2 \\
1 & 0 & 1 & 1/2 & 1/2 & 1/2 & 1 \\
1/2 & 1 & 0 & 1 & 1/2 & 1/2 & 1 \\
1/3 & 1/2 & 1 & 0 & 1 & 1/2 & 1/2 \\
1/3 & 1/2 & 1/2 & 1 & 0 & 1 & 1 \\
1/3 & 1/2 & 1/2 & 1/2 & 1 & 0 & 1 \\
1/2 & 1 & 1 & 1/2 & 1 & 1 & 0
\end{bmatrix}
$$

The edge-distance-edge-degree matrix is nonsymmetric similarly to the vertex-distance-vertex degree matrix for $q \neq r$. This is illustrated for $T_2$ (see Figure 3.1) with the following parameters: $p = 1$, $q = 1$ and $r = 2$. The line graph of $T_2$ is shown in Figure 4.10.

$$
{}^e\mathbf{D}^e\mathbf{d}(1,1,2,T_2) =
\begin{bmatrix}
0 & 32 & 36 & 24 & 8 & 16 & 8 \\
16 & 0 & 36 & 32 & 12 & 16 & 16 \\
24 & 48 & 0 & 12 & 6 & 12 & 24 \\
24 & 64 & 18 & 0 & 2 & 16 & 24 \\
16 & 48 & 8 & 4 & 0 & 12 & 16 \\
16 & 32 & 18 & 16 & 6 & 0 & 16 \\
8 & 32 & 36 & 24 & 8 & 16 & 0
\end{bmatrix}
$$

The distance-degree matrices can be used to generate the distance-degree descriptors (Ivanciuc, 1989; Janežič et al., 2007; Gutman, 2013).

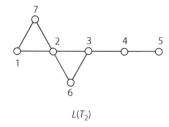

$L(T_2)$

**FIGURE 4.10**   The labeled line graph $L(T_2)$ of the tree $T_2$.

## 4.29   THE RESISTANCE-DISTANCE MATRIX

The *resistance-distance matrix* of a vertex-labeled connected graph $G$, denoted by $\Omega$, is a real symmetric $V \times V$ matrix defined as (Klein and Randić, 1993)

$$[\Omega]_{ij} = \begin{cases} \omega_{ij} & \text{if } i \neq j \\ 0 & \text{otherwise} \end{cases} \tag{4.44}$$

where $\omega_{ij}$ is the resistance-distance between vertices $i$ and $j$.

The resistance-distance matrix was derived on the basis of the theory of resistive electrical networks (Kirchhoff, 1847; Edminster, 1965). An electrical network can be represented by a connected graph $G$ in which the vertices of $G$ correspond to junctions in the electrical network and edges of $G$ correspond to unit resistors (Harary, 1971). The effective resistance-distance between the pairs of vertices is then a graph-theoretical distance, hence the term *resistance-distance*. If a graph is acyclic, then $[\Omega]_{ij}$ is the sum of resistances (each being equal to 1) along the path connecting vertices $i$ and $j$. On the other hand, if a graph contains cycles, Kirchhoff's laws should be employed to obtain $[\Omega]_{ij}$. Properties of the resistance-distance matrix have been studied by several authors (Klein and Randić, 1993; Bonchev et al., 1994; Klein and Zhu, 1998; Ivanciuc and Ivanciuc, 1999; Lukovits et al., 1999, 2000; Fowler, 2002; Klein, 2002; Gutman et al., 2003; Nikolić et al., 2009; Xiao and Gutman, 2003; Zhou and Trinajstić, 2009a; Yang, 2014).

As an example, the resistance-distance matrix of the vertex-labeled graph $G_1$ (see structure $A$ in Figure 2.1) is given below:

$$\Omega(G_1) = \begin{bmatrix} 0 & 1 & 2 & 11/4 & 3 & 11/4 & 15/4 \\ 1 & 0 & 1 & 7/4 & 2 & 7/4 & 11/4 \\ 2 & 1 & 0 & 3/4 & 1 & 3/4 & 7/4 \\ 11/4 & 7/4 & 3/4 & 0 & 3/4 & 1 & 2 \\ 3 & 2 & 1 & 3/4 & 0 & 3/4 & 7/4 \\ 11/4 & 7/4 & 3/4 & 1 & 3/4 & 0 & 1 \\ 15/4 & 11/4 & 7/4 & 2 & 7/4 & 1 & 0 \end{bmatrix}$$

An algorithm based on the Laplacian matrix has been proposed for efficacious computing of the resistance-distance matrix for connected graphs (Babić et al., 2002). This computational algorithm consists of the following steps:

1. Set up a connected graph $G$ with $V$ vertices.
2. Construct the Laplacian matrix $\mathbf{L}$ for $G$.
3. Set up an auxilliary matrix $\mathbf{\Phi}$ of $G$. Matrix $\mathbf{\Phi}$ is a $V \times V$ matrix, all of whose elements are equal to 1.

4. Construct the sum-matrix $\xi = [L + x \, \Phi/V]$, with $x$ having a nonzero arbitrary value bigger than 0. For the simple graphs, the value of $x$ is taken to be unity. For the weighted graphs, the value of $x$ differs from unity.
5. Compute the inverse of the sum-matrix $\xi' = 1/[L + x \, \Phi/V]$. The inverse is nonsingular for connected graphs.
6. Compute the resistance-distance matrix $\Omega$ using the elements of the $\xi'$ matrix: $[\Omega]_{ij} = [\xi']_{ii} - 2 \, [\xi']_{ij} + [\xi']_{jj}$.

Application of the algorithm to $G_1$ (see structure $A$ in Figure 2.1) is presented in Table 4.1.

---

**TABLE 4.1**

**Application of the Algorithm for Computing the Resistance-Distance Matrix of a Simple Graph**

1. Graph $G_1$ (see structure $A$ in Figure 2.1)
2. The Laplacian matrix $L$ of $G_1$ (see Section 2.20)
3. The auxilliary matrix $\Phi$ of $G_1$:

$$\Phi(G_1) = \begin{bmatrix} 1 & 1 & 1 & 1 & 1 & 1 & 1 \\ 1 & 1 & 1 & 1 & 1 & 1 & 1 \\ 1 & 1 & 1 & 1 & 1 & 1 & 1 \\ 1 & 1 & 1 & 1 & 1 & 1 & 1 \\ 1 & 1 & 1 & 1 & 1 & 1 & 1 \\ 1 & 1 & 1 & 1 & 1 & 1 & 1 \\ 1 & 1 & 1 & 1 & 1 & 1 & 1 \end{bmatrix}$$

4. The sum-matrix $\xi$ of $G_1$:

$$\xi(G_1) = 1/7 \begin{bmatrix} 8 & -6 & 1 & 1 & 1 & 1 & 1 \\ -6 & 15 & -6 & 1 & 1 & 1 & 1 \\ 1 & -6 & 22 & -6 & 1 & -6 & 1 \\ 1 & 1 & -6 & 15 & -6 & 1 & 1 \\ 1 & 1 & 1 & -6 & 15 & -6 & 1 \\ 1 & -6 & 1 & 1 & -6 & 22 & -6 \\ 1 & 1 & 1 & 1 & 1 & -6 & 8 \end{bmatrix}$$

5. The inverse sum-matrix $\xi'$ of $G_1$:

$$\xi'(G_1) = \begin{bmatrix} 1.59 & 0.73 & 0.02 & -0.23 & -0.34 & -0.31 & -0.45 \\ 0.73 & 0.87 & 0.16 & -0.09 & -0.20 & -0.16 & -0.31 \\ 0.02 & 0.16 & 0.44 & 0.19 & 0.09 & 0.012 & -0.02 \\ -0.23 & -0.09 & 0.19 & 0.69 & 0.34 & 0.12 & -0.02 \\ -0.34 & -0.20 & 0.09 & 0.34 & 0.73 & 0.27 & 0.12 \\ -0.31 & -0.16 & 0.12 & 0.12 & 0.27 & 0.55 & 0.41 \\ -0.45 & -0.31 & -0.02 & -0.02 & 0.12 & 0.41 & 1.27 \end{bmatrix}$$

6. The resistance-distance matrix $\Omega$ of $G_1$ (see this matrix above)

The Wiener-like distance index, named the Kirchhoff index (Bonchev et al., 1994; Gutman et al., 2003; Zhou and Trinajstić, 2008, 2009b), is based on the resistance-distance matrix. However, it has been elegantly demonstrated (Gutman and Mohar, 1996) that the quasi-Wiener index (Mohar et al., 1993; Gutman et al., 1994; Marković et al., 1995) and the Kirchhoff index are *identical* topological indices.

## 4.30  DISTANCE/DISTANCE MATRICES

The *distance/distance matrices*, denoted by **D/D**, have been introduced by Randić et al. (1994; Randić and Krilov, 1999) and discussed by Todeschini and Consonni (2000, 2009). These matrices are defined in terms of the geometric or topographic distances $g(i, j)$ and the graph-theoretical (topological) distances $l(i, j)$. Thus, they unify the topological and topographic (geometric) information on the structure of a given molecule.

We consider here two kinds of distance/distance matrices: the topographic distance/topological distance matrix and the corresponding reciprocal matrix. We denote the first matrix as $^g\mathbf{D}/^t\mathbf{D}$ and the second as $^t\mathbf{D}/^g\mathbf{D}$.

The topographic distance/topological distance matrix is an unsymmetric $V \times V$ matrix, in which the upper matrix-triangle represents the same part of the corresponding topographic matrix, and the lower matrix-triangle the same part of vertex-distance matrix (defined in Section 4.1):

$$\left[^g\mathbf{D}/^t\mathbf{D}\right]_{ij} = \begin{cases} g(i,j) & \text{if } i > j \\ 0 & \text{if } i = j \\ l(i,j) & \text{if } i < j \end{cases} \tag{4.45}$$

The topological distance matrix of $T_2$ (see Figure 2.19) $^t\mathbf{D}$ is given by

$$^t\mathbf{D}(T_2) = \begin{bmatrix} 0 & 1 & 2 & 3 & 4 & 5 & 3 & 2 \\ 1 & 0 & 1 & 2 & 3 & 4 & 2 & 1 \\ 2 & 1 & 0 & 1 & 2 & 3 & 1 & 2 \\ 3 & 2 & 1 & 0 & 1 & 2 & 2 & 3 \\ 4 & 3 & 2 & 1 & 0 & 1 & 3 & 4 \\ 5 & 4 & 3 & 2 & 1 & 0 & 4 & 5 \\ 3 & 2 & 1 & 2 & 3 & 4 & 0 & 3 \\ 2 & 1 & 2 & 3 & 4 & 5 & 3 & 0 \end{bmatrix}$$

The geometric distance matrices are based on the actual molecular geometry of a molecule in 3-D space (Bogdanov et al., 1989, 1990; Balasubramanian, 1990; Nikolić et al., 1991), while the topographic matrices are based on standardized bond angles and bond lengths (Randić et al., 1994). They can be derived by embedding

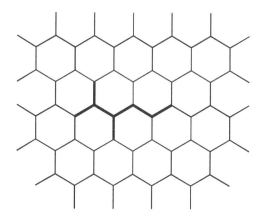

**FIGURE 4.11**   Branched tree $T_2$ embedded on a graphite grid.

a graph on a regular 2-D or 3-D grid. The embedding of $T_2$ (see Figure 2.19) on a graphite (honeycomb) grid is shown in Figure 4.11.

The topographic distance matrix $^g\mathbf{D}$ is defined as (Randić, 1988)

$$\left[ ^g\mathbf{D} \right]_{ij} = \begin{cases} g(i,j) & \text{if } i \neq j \\ 0 & \text{otherwise} \end{cases} \tag{4.46}$$

where $g(i,j)$ is the topographic distance.

The elements of this matrix are computed by taking the edge-distance to be unity and using the plane geometry. It should be noted that the structures of topographic matrices depend on the embedding—different embeddings of the same graph result in different matrices. As a result, structural invariants derived from such matrices depend on the assumed conformation of a graph (Randić, 1988).

The topographic distance matrix $^g\mathbf{D}$ of $T_2$ (see Figure 2.19) is given by

$$^g\mathbf{D}(T_2) = \begin{bmatrix} 0 & 1 & \sqrt{3} & \sqrt{7} & 2\sqrt{3} & \sqrt{19} & 2 & \sqrt{3} \\ 1 & 0 & 1 & \sqrt{3} & \sqrt{7} & 2\sqrt{3} & \sqrt{3} & 1 \\ \sqrt{3} & 1 & 0 & 1 & \sqrt{3} & \sqrt{7} & 1 & \sqrt{3} \\ \sqrt{7} & \sqrt{3} & 1 & 0 & 1 & \sqrt{3} & \sqrt{3} & 2 \\ 2\sqrt{3} & \sqrt{7} & \sqrt{3} & 1 & 0 & 1 & 2 & 3 \\ \sqrt{19} & 2\sqrt{3} & \sqrt{7} & \sqrt{3} & 1 & 0 & 3 & \sqrt{13} \\ 2 & \sqrt{3} & 1 & \sqrt{3} & 2 & 3 & 0 & \sqrt{7} \\ \sqrt{3} & 1 & \sqrt{3} & 1/2 & 3 & \sqrt{13} & \sqrt{7} & 0 \end{bmatrix}$$

The topographic distance/topological distance matrix $^g\mathbf{D}/^t\mathbf{D}$ of $T_2$ is given by

$$
^g\mathbf{D}/^t\mathbf{D}(T_2) =
\begin{bmatrix}
0 & 1 & \sqrt{3} & \sqrt{7} & 2\sqrt{3} & \sqrt{19} & 2 & \sqrt{3} \\
1 & 0 & 1 & \sqrt{3} & \sqrt{7} & 2\sqrt{3} & \sqrt{3} & 1 \\
2 & 1 & 0 & 1 & \sqrt{3} & \sqrt{7} & 1 & \sqrt{3} \\
3 & 2 & 1 & 0 & 1 & \sqrt{3} & \sqrt{3} & 2 \\
4 & 3 & 2 & 1 & 0 & 1 & 2 & 3 \\
5 & 4 & 3 & 2 & 1 & 0 & 3 & \sqrt{13} \\
3 & 2 & 1 & 2 & 3 & 4 & 0 & \sqrt{7} \\
2 & 1 & 2 & 3 & 4 & 5 & 3 & 0
\end{bmatrix}
$$

The normalized Perron root (the first eigenvalue) (Horn and Johnson, 1985) of such matrices for linear structures appears to be an index of molecular folding (Randić and Krilov, 1999).

The topological distance/topographic distance matrix is an unsymmetric $V \times V$ matrix, in which the upper matrix-triangle represents the same part of the corresponding vertex-distance matrix, and the lower matrix-triangle the same part of topographic matrix

$$
\left[ {}^t\mathbf{D}/{}^g\mathbf{D} \right]_{ij} =
\begin{cases}
g(i,j) & \text{if } i < j \\
0 & \text{if } i = j \\
l(i,j) & \text{if } i > j
\end{cases}
\tag{4.47}
$$

The topological distance/topographic distance matrix $^t\mathbf{D}/^g\mathbf{D}$ of $T_2$ is given by

$$
^t\mathbf{D}/^g\mathbf{D}(T_2) =
\begin{bmatrix}
0 & 1 & 2 & 3 & 4 & 5 & 3 & 2 \\
1 & 0 & 1 & 2 & 3 & 4 & 2 & 1 \\
\sqrt{3} & 1 & 0 & 1 & 2 & 3 & 1 & 2 \\
\sqrt{7} & \sqrt{3} & 1 & 0 & 1 & 2 & 2 & 3 \\
2\sqrt{3} & \sqrt{7} & \sqrt{3} & 1 & 0 & 1 & 3 & 4 \\
\sqrt{19} & 2\sqrt{3} & \sqrt{7} & \sqrt{3} & 1 & 0 & 4 & 5 \\
2 & \sqrt{3} & 1 & \sqrt{3} & 2 & 3 & 0 & 3 \\
\sqrt{3} & 1 & \sqrt{3} & 2 & 3 & \sqrt{13} & \sqrt{7} & 0
\end{bmatrix}
$$

A number of graph invariants can be obtained from the distance/distance matrices made up from geometric, topographic, and topological matrices (Todeschini and Consonni, 2009).

## 4.31   THE COMMON VERTEX MATRIX

Randić and his coworkers (Randić et al., 2013a) introduced a novel matrix representation of graphs, called *common vertex matrix*, denoted by **CVM**. It is defined as follows:

$$[\mathbf{CVM}]_{ij} = \begin{cases} \Lambda & \text{the number of vertices at equal distance} \\ & \text{from vertices } i \text{ and } j \\ 0 & \text{if } i = j \end{cases} \tag{4.48}$$

We illustrate the use of this matrix for $T_2$, shown in Figure 4.12. In this figure, the vertices in $T_2$ are labeled so that the corresponding **CVM** matrix assumes the block-diagonal form. Since trees belong to the class of bipartite graphs (Trinajstić, 1983, 1992), their vertices can be split into two classes denoted by small circles and black dots, respectively.

The corresponding common vertex matrix **CVM** of $T_2$ (see Figure 2.19) is shown below:

$$\mathbf{CVM}(T_2) = \begin{bmatrix} 0 & 2 & 2 & 6 & 0 & 0 & 0 & 0 \\ 2 & 0 & 1 & 2 & 0 & 0 & 0 & 0 \\ 2 & 1 & 0 & 2 & 0 & 0 & 0 & 0 \\ 6 & 2 & 2 & 0 & 0 & 0 & 0 & 0 \\ 0 & 0 & 0 & 0 & 0 & 2 & 1 & 4 \\ 0 & 0 & 0 & 0 & 2 & 0 & 1 & 4 \\ 0 & 0 & 0 & 0 & 1 & 1 & 0 & 1 \\ 0 & 0 & 0 & 0 & 4 & 4 & 1 & 0 \end{bmatrix}$$

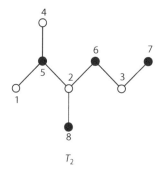

$T_2$

**FIGURE 4.12** The conveniently labeled tree $T_2$ representing the carbon skeleton of 2,3-dimethylhexane.

Randić et al. (2013a) illustrated on smaller alkanes that **CVM** is sensitive to branching, and that the ordered row sums of **CVM** may facilitate solving a graph isomerism problem for acyclic graphs. Randić and his coworkers (Randić et al., 2013b) used the common vertex matrix for novel characterization of the central vertex or vertices in acyclic and cyclic graphs.

## REFERENCES

D. Amić and N. Trinajstić, On the detour matrix, *Croat. Chem. Acta* 68 (1995) 53–62.

D. Babić, D.J. Klein, I. Lukovits, S. Nikolić, and N. Trinajstić, Resistance distance matrix: A computional algorithm and its application, *Int. J. Quantum Chem.* 90 (2002) 166–176.

A.T. Balaban, Topological indices based on topological distances in molecular graphs, *Pure Appl. Chem.* 55 (1983) 199–206.

A.T. Balaban, Chemical graphs. 48. Topological index *J* for heteroatom-containing molecules taking in to account periodicities of element properties, *MATCH Commun. Math. Comput. Chem.* 21 (1986) 115–122.

A.T. Balaban, Highly discriminating distance-based topological index, *Chem. Phys. Lett.* 89 (1989) 399–404.

A.T. Balaban and O. Ivanciuc, FORTRAN 77 computer program for calculating the topological index *J* for molecules containing heteroatoms, in *MATH/CHEM/COMP 1988*, ed. A. Graovac, Elsevier, Amsterdam, 1989, pp. 193–211.

A.T. Balaban, D. Mills, O. Ivanciuc, and S.C. Basak, Reverse Wiener indices, *Croat. Chem. Acta* 73 (2000) 923–941.

T.S. Balaban, P.A. Filip, and O. Ivanciuc, Computer generation of acyclic graphs based on local vertex invariants and topological indices. Derived canonical labelling and coding of trees and alkanes, *J. Math. Chem.* 11 (1992) 79–105.

K. Balasubramanian, Geometry-dependent characteristic polynomials of molecular structures, *Chem. Phys. Lett.* 169 (1990) 224–228.

M. Barysz, G. Jashari, R.S. Lall, V.K. Srivastava, and N. Trinajstić, On the distance matrix of molecules containing heteroatoms, in *Chemical applications of topology and graph theory*, ed. R.B. King, Elsevier, Amsterdam, 1983, pp. 222–230.

N.L. Biggs, E.K. Lloyd, and R.J. Wilson, *Graph theory 1736–1936*, reprinted with corrections, Clarendon Press, Oxford, 1977; see chap. 8, pp. 131–157.

B. Bogdanov, S. Nikolić, and N. Trinajstić, On the three-dimensional Wiener number, *J. Math. Chem.* 3 (1989) 299–309.

B. Bogdanov, S. Nikolić, and N. Trinajstić, On the three-dimensional Wiener number. A comment, *J. Math. Chem.* 5 (1990) 305–306.

D. Bonchev, A.T. Balaban, X. Liu, and D.J. Klein, Molecular cyclicity and centricity of polycyclic graphs. I. Cyclicity based on resistance distances or reciprocal distances, *Int. J. Quantum Chem.* 50 (1994) 1–20.

D. Bonchev and D.J. Klein, On the Wiener number of thorn trees, stars, rings and rods, *Croat. Chem. Acta* 75 (2002) 613–620.

D. Bonchev and N. Trinajstić, Information theory, distance matrix and molecular branching, *J. Chem. Phys.* 67 (1977) 4517–4533.

F. Buckley and F. Harary, *Distance in graphs*, Addison-Wesley, Reading, MA, 1990.

V. Consonni and R. Todeschini, Multivariate analysis of molecular descriptors, in *Statistical modelling of molecular descriptors in QSAR/QSPR*, ed. M. Dehmer, K. Varmuza, and D. Bonchev, Wiley-Blackwell, Weinheim, 2012, pp. 111–147.

M.V. Diudea, Walk numbers $^eW_M$: Wiener-type numbers of higher rank, *J. Chem. Inf. Comput. Sci.* 36 (1996a) 535–540.

M.V. Diudea, Wiener and hyper-Wiener numbers in a single matrix, *J. Chem. Inf. Comput. Sci.* 36 (1996b) 833–836.

M.V. Diudea, Indices of reciprocal properties or Harary indices, *J. Chem. Inf. Comput. Sci.* 37 (1997) 292–299.

M.V. Diudea, M.S. Florescu, and P.V. Khadikar, *Molecular topology and its applications*, EfiCon Press, Bucharest, 2006.

A.A. Dobrynin, R. Entringer, and I. Gutman, Wiener index of trees: Theory and application, *Acta Appl. Math.* 66 (2000) 211–249.

H.E. Dudeney, *Amusements in mathematics*, Nelson, London, 1917.

J.A. Edminster, *Electric circuits*, McGraw-Hill, New York, 1965.

P.W. Fowler, Resistance distances in fullerene graphs, *Croat. Chem. Acta* 75 (2002) 401–408.

J. Gálvez, R. Garcia, M.T. Salabert, and R. Soler, Charge indexes: New topological descriptors, *J. Chem. Inf. Comput. Sci. 34* (1994) 520–525.

J. Gálvez, R. Garcia-Domenech, J.V. De Julián-Ortiz, and R. Soler, Topological approach to drug design, *J. Chem. Inf. Comput. Sci.* 35 (1995) 272–284.

A. Graovac, O.E. Polansky, N. Trinajstić, and N. Tyutyulkov, Graph theory in chemistry. II. Graph-theoretical description of heteroconjugated molecules, *Z. Naturforsch.* 30a (1975) 1696–1699.

I. Gutman, A formula for the Wiener number of trees and its extension to graph-containing cycles, *Graph Theory Notes NY* 27 (1994) 9–15.

I. Gutman, A new hyper-Wiener index, *Croat. Chem. Acta* 77 (2004) 61–64.

I. Gutman, Degree-based topological indices, *Croat. Chem. Acta* 86 (2013) 351–261.

I. Gutman, S.-L. Lee, C.-H. Chu, and Y.-L. Luo, Chemical applications of the Laplacian spectrum of molecular graphs: Studies of the Wiener number, *Ind. J. Chem.* 33A (1994) 603–608.

I. Gutman, W. Linert, I. Lukovits, and Ž. Tomović, On the multiplicative Wiener index and its possible chemical applications, *Monat. Chem.* 131 (2000a) 421–427.

I. Gutman, W. Linert, I. Lukovits, and Ž. Tomović, The multiplicative version of the index, *J. Chem. Inf. Comput. Sci.* 40 (2000b) 113–116.

I. Gutman and B. Mohar, The quasi-Wiener and the Kirchhoff indices coincide, *J. Chem. Inf. Comput. Sci.* 36 (1996) 982–985.

I. Gutman and O.E. Polansky, *Mathematical concepts in organic chemistry*, Springer, Berlin, 1986.

I. Gutman, D. Vidović, and B. Furtula, Chemical applications of the Laplacian spectrum. VII. Studies of the Wiener and Kirchhoff indices, *Ind. J. Chem.* 42A (2003) 1272–1278.

I. Gutman, D. Vukičević, and J. Žerovnik, A class of modified Wiener indices, *Croat. Chem. Acta* 77 (2004) 103–109.

I. Gutman and J. Žerovnik, Corroborating a modification of the Wiener number, *Croat. Chem. Acta* 75 (2002) 603–612.

I. Gutman, Y.N. Yeh, S.L. Lee, and Y.L. Luo, Some recent results in the theory of the Wiener number, *Indian J. Chem.* 32A (1993) 651–661.

S.L. Hakimi and S.S. Yau, Distance matrix of a graph and its realizability, *Q. Appl. Math.* 12 (1965) 305–317.

F. Harary, Combinatorial problems in graphical enumeration, in *Applied combinatorial mathematics*, ed. E.F. Beckenbach, Wiley, New York, 1964, pp. 185–220.

F. Harary, Applications of Pólya theorem to permutations groups, in *A seminar on graph theory*, ed. F. Haray and L.W. Beineke, Holt, Rinehart and Winston, New York, 1967, pp. 25–33.

F. Harary, *Graph theory*, 2nd printing, Addison-Wesley, Reading, MA, 1971.

R.A. Horn and C.R. Johnson, *Matrix analysis*, University Press, Cambridge, 1985.

H. Hosoya, Topological index. A newly proposed quantity characterizing the topological nature of structural isomers of saturated hydrocarbons, *Bull. Chem. Soc. Jpn.* 44 (1971) 2332–2339.

H. Hosoya, Distance polynomial and the related counting polynomials, *Croat. Chem. Acta* 86 (2013) 443–451.

H. Hosoya, U. Nagashima, and S. Hyugaji, Topological twin graphs. Smallest pair of isospectral polyhedral graphs with eight vertices, *J. Chem. Inf. Comput. Sci.* 34 (1994) 428–431.

H. Hosoya, K. Ohta, and M. Satomi, Topological twin graphs. II. Isospectral polyhedral graphs with nine and ten vertices, *MATCH Commun. Math. Comput. Chem.* 44 (2001) 183–200.

O. Ivanciuc, Design of topological indices. Part 1. Definition of a vertex topological index in the case of 4-trees, *Rev Roum. Chim.* 34 (1989) 1361–1368.

O. Ivanciuc, Design of topological indices. Part 11. Distance-valency matrices and derived molecular graph descriptors, *Rev Roum. Chim.* 44 (1999) 519–528.

O. Ivanciuc, Design of topological indices. Part 12. Parameters for vertex- and edge-weighted molecular graphs, *Rev. Roum. Chim.* 45 (2000a) 289–301.

O. Ivanciuc, QSAR comparative study of Wiener descriptors for weighted molecular graphs, *J. Chem. Inf. Comput. Sci.* 40 (2000b) 1412–1422.

O. Ivanciuc, Design of topological indices. Part 14. Distance-valency matrices and structural descriptors for vertex- and edge-weighted molecular graphs, *Rev Roum. Chim.* 45 (2000c) 587–596.

O. Ivanciuc and A.T. Balaban, Design of topological indices. Part 8. Path matrices and derived molecular graph invariants, *MATCH Commun. Math. Comput. Chem.* 30 (1994) 141–152.

O. Ivanciuc, T.-S. Balaban, and A.T. Balaban, Design of topological indices. Part 4. Reciprocal distance matrix, related local vertex invariants and topological indices, *J. Math. Chem.* 12 (1993) 309–318.

O. Ivanciuc and T. Ivanciuc, Matrices and structural descriptors computed from molecular graph distances, in *Topological indices and related descriptors in QSAR and QSPR*, ed. J. Devillers and A.T. Balaban, Gordon & Breach, Amsterdam, 1999, pp. 221–277.

O. Ivanciuc, T. Ivanciuc, and A.T. Balaban, Design of topological indices. Part 10. Parameters based on electronegativity and covalent radius for the computation of molecular graph descriptors for heteroatom-containing molecules, *J. Chem. Inf. Comput. Sci.* 38 (1998) 395–401.

O. Ivanciuc, T. Ivanciuc, and A.T. Balaban, Vertex- and edge-weighted molecular graphs and derived structural descriptors, in *Topological indices and related descriptors in QSAR and QSPR*, ed. J. Devillers and A.T. Balaban, Gordon & Breach, Amsterdam, 1999, pp. 169–230.

O. Ivanciuc, T. Ivanciuc, and A.T. Balaban, The complementary distance matrix, a new molecular graph metric, *ACH-Model. Chem.* 137 (2000) 57–82.

O. Ivanciuc and D.J. Klein, Building-block computation of Wiener-type indices for virtual screening of combinatorial libraries, *Croat. Chem. Acta* 75 (2002) 577–601.

D. Janežič, B. Lučić, A. Miličević, S. Nikolić, N. Trinajstić, and D. Vukičević, Hosoya matrices as the numerical realization of graphical matrices and derived structural descriptors, *Croat. Chem. Acta* 80 (2007) 271–276.

P.E. John and M.V. Diudea, Wiener index of zig-zag polyhex nanotubes, *Croat. Chem. Acta* 77 (2004) 127–132.

G. Kirchhoff, Über die Auflösung der Gleichungen, auf welche man bei der Untersuchung der linearen Verteilung galvanischer Ströme gefürt wird, *Ann. Phys. Chem.* 72 (1847) 497–508; English translation in J.B. O'Toole, On the solution of the equations obtained from the investigation of the linear distribution of galvanic currents, *Trans IRE* CT-5 (1958) 4–7.

D.J. Klein, Resistance-distance sum rules, *Croat. Chem. Acta* 75 (2002) 633–649.

D.J. Klein, I. Lukovits, and I. Gutman, On the definition of the hyper-Wiener index for cycle-containing structures, *J. Chem. Inf. Comput. Sci.* 35 (1995) 50–52.

D.J. Klein and M. Randić, Resistance distance, *J. Math. Chem.* 12 (1993) 81–95.

D.J. Klein and H.-Y. Zhu, Distances and volumina for graphs, *J. Math. Chem.* 23 (1998) 179–195.

J.K. Kruskal Jr., On the shortest spanning subtree of a graph and the traveling salesman, *Proc. Am. Math. Soc.* 7 (1956) 48–50.

K. Kuratowski, Sur le problème des courbes gauches en topologie, *Fund. Math.* 15 (1930) 271–283.

C. Liang and K. Mislow, Topological chirality of proteins, *J. Am. Chem. Soc.* 116 (1994) 3588–3592.

S. Lloyd Jr., *Sam Lloyd and his puzzles*, Barse, New York, 1928.

B. Lučić, I. Lukovits, S. Nikolić, and N. Trinajstić, Distance-related indexes in the quantitative structure-property relationship modeling, *J. Chem. Inf. Comput. Sci.* 41 (2001) 527–535.

B. Lučić, A. Miličević, S. Nikolić, and N. Trinajstić, Harary index—Twelve years later, *Croat. Chem. Acta* 75 (2002) 847–868.

B. Lučić, A. Miličević, S. Nikolić, and N. Trinajstić, On variable Wiener index, *Ind. J. Chem.* 42A (2003) 1279–1282.

B. Lučić, I. Sović, D. Plavšić, and N. Trinajstić, Harary matrices: Definitions, properties and applications, in *Distance in molecular graphs—Applications*, ed. I. Gutman and B. Furtula, University of Kragujevac, Kragujevac, Serbia, 2012, pp. 3–26.

I. Lukovits, The detour index, *Croat. Chem. Acta* 69 (1996) 873–882.

I. Lukovits and W. Linert, A novel definition of the hyper-Wiener index for cycles, *J. Chem. Inf. Comput. Sci.* 34 (1994) 899–902.

I. Lukovits, S. Nikolić, and N. Trinajstić, Resistance distance in regular graphs, *Int. J. Quantum Chem.* 71 (1999) 217–225.

I. Lukovits, S. Nikolić, and N. Trinajstić, Note on the resistance distances in the dodecahedron, *Croat. Chem. Acta* 73 (2000) 957–967.

I. Lukovits and M. Razinger, On calculation of the detour index, *J. Chem. Inf. Comput. Sci.* 37 (1997) 283–286.

R.B. Mallion, A.J. Schwenk, and N. Trinajstić, A graphical study of heteroconjugated molecules, *Croat. Chem. Acta* 46 (1974a) 171–182.

R.B. Mallion, A.J. Schwenk, and N. Trinajstić, On the characteristic polynomial of a rooted graph, in *Recent advances in graph theory*, ed. M. Fiedler, Academia, Prague, 1975, pp. 345–350.

R.B. Mallion, N. Trinajstić, and A.J. Schwenk, Graph theory in chemistry. Generalization of Sachs' formula, *Z. Naturforsch.* 29a (1974b) 1481–1484.

S. Marković, I. Gutman, and Ž. Bančević, Correlation between Wiener and quasi-Wiener indices in benzenoid hydrocarbons, *J. Serb. Chem. Soc.* 60 (1995) 633–636.

Z. Mihalić and N. Trinajstić, A graph-theoretical approach to structure-property relationships, *J. Chem. Educ.* 69 (1992) 701–712.

Z. Mihalić, D. Veljan, D. Amić, S. Nikolić, D. Plavšić, and N. Trinajstić, The distance matrix in chemistry, *J. Math. Chem.* 11 (1992) 223–258.

B. Mohar, D. Babić, and N. Trinajstić, A novel definition of the Wiener index for trees, *J. Chem. Inf. Comput. Sci.* 33 (1993) 153–154.

W.R. Müller, K. Szymanski, J. von Knop, and N. Trinajstić, An algorithm for construction of the molecular distance matrix, *J. Comput. Chem.* 8 (1987) 170–173.

J.K. Nagle, Atomic polarizability and electronegativity, *J. Am. Chem. Soc.* 112 (1990) 4741–747.

S. Nikolić, D. Plavšić, and N. Trinajstić, On the Balaban-like topological indices, *MATCH Commun. Math. Comput. Chem.* 44 (2001a) 361–386.

S. Nikolić, N. Trinajstić, A. Jurić, and Z. Mihalić, The detour matrix and the detour index of weighted graphs, *Croat. Chem. Acta* 69 (1996) 1577–1591.

S. Nikolić, N. Trinajstić, J. von Knop, W.R. Müller, and K. Szymanski, On the concept of the weighted spanning tree of dualist, *J. Math. Chem.* 4 (1990) 357–375.

S. Nikolić, N. Trinajstić, and Z. Mihalić, The Wiener index: Development and application, *Croat. Chem. Acta* 68 (1995) 105–129.

S. Nikolić, N. Trinajstić, and Z. Mihalić, The detour matrix and the detour index, in *Topological indices and related descriptors in QSAR and QSPR*, ed. J. Devillers and A.T. Balaban, Gordon & Breach, Amsterdam, 1999, pp. 279–306.

S. Nikolić, N. Trinajstić, Z. Mihalić, and S. Carter, On the geometric-distance matrix and the corresponding structural invariants of molecular systems, *Chem. Phys. Lett.* 179 (1991) 21–28.

S. Nikolić, N. Trinajstić, and M. Randić, Wiener index revisited, *Chem. Phys. Lett.* 333 (2001b) 319–321.

S. Nikolić, N. Trinajstić, and B. Zhou, On the eigenvalues of the ordinary and reciprocal resistance-distance matrix, in *Computational methods in science and engineering*, Vol. I, ed. G. Maroulis and T.E. Simos, American Institute of Physics, Melville, NY, 2009, pp. 205–214.

A.N. Patrinos and S.L. Hakimi, The distance matrix of a graph and its tree realization, *Q. Appl. Math.* 30 (1973) 255–269.

D. Plavšić, S. Nikolić, N. Trinajstić, and Z. Mihalić, On Harary index for the characterization of chemical graphs, *J. Math. Chem.* 12 (1993) 235–250.

M. Randić, Molecular topographic descriptors, in *MATH/CHEM/COMP 1987*, ed. R.C. Lacher, Elsevier, 1988, pp. 101–108.

M. Randić, Novel molecular descriptor for structure-property studies, *Chem. Phys. Lett.* 211 (1993) 478–483.

M. Randić, $D_{MAX}$—Matrix of dominant distances in a graph, *MATCH Commun. Math. Comput. Chem.* 70 (2013) 221–238.

M. Randić, L.M. DeAlba, and F.E. Harris, Graphs with the same detour matrix, *Croat. Chem. Acta* 71 (1998) 53–68.

M. Randić, V. Katović, and N. Trinajstić, Symmetry properties of chemical graphs. VII. Enantiomers of a tetragonal-pyramidal rearrangement, in *Symmetries and properties of non-rigid molecules: A comprehensive survey*, ed. J. Maruani and J. Serre, Elsevier, Amsterdam, 1983, pp. 399–408.

M. Randić, A.F. Kleiner, and L.M. DeAlba, Distance/distance matrices, *J. Chem. Inf. Comput. Sci.* 34 (1994) 277–286.

M. Randić and G. Krilov, On a characterization of the folding of proteins, *Int. J. Quantum Chem.* 75 (1999) 1017–1026.

M. Randić, M. Novič, and D. Plavšić, Common vertex matrix: A novel characterization of molecular graphs by counting, *J. Comput. Chem.* 34 (2013a) 1409–1419.

M. Randić, M. Novič, M. Vračko, and D. Plavšić, On the centrality of vertices in molecular graphs, *J. Comput. Chem.* 34 (2013b) 2514–2523.

M. Randić and M. Pompe, The variable molecular descriptors based on distance-related indices, *J. Chem. Inf. Comput. Sci.* 41 (2001) 575–581.

M. Randić, A. Sabljić, S. Nikolić, and N. Trinajstić, A rational selection of graph-theoretical indices in the QSAR, *Int. J. Quantum Chem. Quantum Biol. Symp.* 15 (1988) 267–285.

M. Randić and J. Zupan, On interpretation of well-known topological indices, *J. Chem. Inf. Comput. Sci.* 41 (2001) 550–560.

D.H. Rouvray, Predicting chemistry from topology, *Sci. Am.* 254 (9) (1986) 40–47.

G. Rücker and C. Rücker, Symmetry-aided computation of the detour matrix and the detour index, *J. Chem. Inf. Comput. Sci.* 38 (1998) 710–714.

R.T. Sanderson, *Polar Covalence*, Academic, New York, 1983.

P.G. Seybold, Topological influences on the carcinogenecity of aromatic hydrocarbons. I. Bay region geometry, *Int. J. Quantum Chem. Quantum Biol. Symp.* 10 (1983) 95–101.

R. Todeschini and V. Consonni, *Handbook of molecular descriptors*, Wiley-VCH, Weinheim, 2000.

R. Todeschini and V. Consonni, *Molecular descriptors for chemoinformatics*, Vols. I and II, Wiley-VCH, Weinheim, 2009.

N. Trinajstić, *Chemical graph theory*, Vols. I and II, CRC, Boca Raton, FL, 1983.

N. Trinajstić, *Chemical graph theory*, 2nd ed., CRC, Boca Raton, FL, 1992.

N. Trinajstić, S. Nikolić, S.C. Basak, and I. Lukovits, Distance indices and their hyper-counterparts: Intercorrelation and use in the structure-property modeling, *SAR QSAR Environ. Res.* 12 (2001) 31–54.

N. Trinajstić, S. Nikolić, B. Lučić, D. Amić, and Z. Mihalić, The detour matrix in chemistry, *J. Chem. Inf. Comput. Sci.* 37 (1997a) 631–638.

N. Trinajstić, S. Nikolić, and Z. Mihalić, On computing the molecular detour matrix, *Int. J. Quantum Chem.* 65 (1997b) 415–419.

H. Wiener, Structural determination of paraffin boiling points, *J. Am. Chem. Soc.* 69 (1947) 17–20.

W. Xiao and I. Gutman, On resistance matrices, *MATCH Commun. Math. Comput. Chem.* 49 (2003) 67–81.

Y. Yang, Relation between resistance distances of a graph and its complement or its contraction, *Croat. Chem. Acta* 87 (2014) 61–68.

B. Zhou and N. Trinajstić, A note on Kirchhoff index, *Chem. Phys. Lett.* 455 (2008) 120–123.

B. Zhou and N. Trinajstić, On reciprocal and reverse Balaban indices, *Croat. Chem. Acta* 82 (2009a) 537–541.

B. Zhou and N. Trinajstić, On resistance-distance and Kirchhoff index, *J. Math. Chem.* 46 (2009b) 283–289.

B. Zhou and N. Trinajstić, Mathematical properties of molecular descriptors based distances, *Croat. Chem. Acta* 83 (2010) 227–242.

# 5 Special Matrices

Most special graph-theoretical matrices have been introduced to serve as the sources for deriving novel classes of molecular descriptors. Many such matrices are available in the literature. We have selected here several of them to demonstrate the process of generating novel graph-theoretical matrices.

## 5.1 ADJACENCY-PLUS-DISTANCE MATRICES

The initial *adjacency-plus-distance matrix*, or *vertex-Schultz matrix*, denoted by $^v\mathbf{SM}$, has been introduced by Schultz (1989) and formalized by Müller et al. (1990). It is defined as the sum of the vertex-adjacency matrix and the vertex-distance matrix:

$$^v\mathbf{SM} = {^v\mathbf{A}} + {^v\mathbf{D}} \tag{5.1}$$

An example of this matrix for the vertex-labeled graph $G_1$ (see structure $A$ in Figure 2.1) is as follows:

$$
^v\mathbf{SM}(G_1) = \begin{bmatrix}
0 & 2 & 2 & 3 & 4 & 3 & 4 \\
2 & 0 & 2 & 2 & 3 & 2 & 3 \\
2 & 2 & 0 & 2 & 2 & 2 & 2 \\
3 & 2 & 2 & 0 & 2 & 2 & 3 \\
4 & 3 & 2 & 2 & 0 & 2 & 2 \\
3 & 2 & 2 & 2 & 2 & 0 & 2 \\
4 & 3 & 2 & 3 & 2 & 2 & 0
\end{bmatrix}
$$

Schultz used the vertex-adjacency-plus-vertex-distance matrix $^v\mathbf{A} + {^v\mathbf{D}}$ to derive what he called the *molecular topological index* (MTI) for alkanes by multiplying the row-matrix containing the vertex-degrees with $^v\mathbf{SM}$, and then summing up the elements of the obtained row-matrix (Schultz, 1989). Later on, the molecular topological index was named the *Schultz index* (Trinajstić, 1992; Mihalić and Trinajstić, 1992; Mihalić et al., 1992a), and this term is now generally accepted (e.g., Devillers and Balaban, 1999; Todeschini and Consonni, 2000, 2009; Diudea et al., 2001; Hong et al., 2012). The Schultz index has found moderate use in the structure-property-activity modeling (e.g., Mihalić and Trinajstić, 1992; Mihalić et al., 1992b; Jurić et al., 1992; Todeschini and Consonni, 2000, 2009). H.P. Schultz and his collaborators, E.B. Schultz and T.P. Schultz, published 12 papers on various aspects of the Schultz index (Todeschini and Consonni, 2009). It has also been shown that the Shultz index and the Wiener index are closely related graph-theoretical invariants for acyclic structures (Klein et al., 1992; Plavšić et al., 1993; Gutman, 1994b; Gutman and Klavžar, 1997). A determinant of the $^v\mathbf{SM}$ matrix has also been used as a molecular descriptor (Schultz et al., 1990; von Knop et al., 1991).

From (5.1), it can easily be seen that various versions of the adjacency matrix and distance matrix may be employed to generate other kinds of $\mathbf{SM}$ matrices and the corresponding versions of Schultz indices. For example, Estrada and Gutman (1996) introduced the edge-adjacency-plus-edge-distance matrix of a graph $G$:

$$^e\mathbf{SM} = {}^e\mathbf{A} + {}^e\mathbf{D} \tag{5.2}$$

This matrix is equal to the vertex-adjacency-plus-vertex-distance matrix of the respective line graph $L(G)$. The $^e\mathbf{SM}$ matrix of $G_1$ or $^v\mathbf{SM}$ of $L(G_1)$ (see Figure 2.12) is as follows:

$$^e\mathbf{SM}[L(G_1)] = \begin{bmatrix} 0 & 2 & 2 & 3 & 3 & 3 & 2 \\ 2 & 0 & 2 & 2 & 2 & 2 & 2 \\ 2 & 2 & 0 & 2 & 2 & 2 & 2 \\ 3 & 2 & 2 & 0 & 2 & 2 & 2 \\ 3 & 2 & 2 & 2 & 0 & 2 & 2 \\ 3 & 2 & 2 & 2 & 2 & 0 & 2 \\ 2 & 2 & 2 & 2 & 2 & 2 & 0 \end{bmatrix}$$

Estrada and Gutman (1996) used this matrix to derive the edge-molecular topological index that we call the edge-Schultz index.

## 5.2   THE DISTANCE-SUM-CONNECTIVITY MATRIX

One can generate the *distance-sum-connectivity matrix*, denoted by $^\delta\chi$, if one substitutes vertex-degrees with the distance-sums (Szymanski et al., 1986) in the formula for the vertex-connectivity matrix, presented in Section 2.13:

$$\left[ {}^\delta\chi \right]_{ij} = \begin{cases} [\delta(i)\delta(j)]^{-1/2} & \text{if vertices } i \text{ and } j \text{ are adjacent} \\ 0 & \text{otherwise} \end{cases} \tag{5.3}$$

where the distance-sum is defined as (Seybold, 1983)

$$\delta(i) = \sum_{j=1}^{V} \left[ {}^v\mathbf{D} \right]_{ij} \tag{5.4}$$

For example, the distance-sums of vertices in $G_1$ (see structure $A$ Figure 2.1), obtained from the corresponding vertex-distance matrix, given in Section 4.1, are as follows (vertex-labels are in parentheses): 17 (1), 12 (2), 9 (3), 12 (4), 13 (5), 10 (6), and

15 (7). The distance-sum-connectivity matrix of $G_1$ (see structure $A$ in Figure 2.1) is a square $7 \times 7$ matrix, given below:

$$^\delta\chi(G_1) = \begin{bmatrix} 0 & 0.070 & 0 & 0 & 0 & 0 & 0 \\ 0.070 & 0 & 0.096 & 0 & 0 & 0 & 0 \\ 0 & 0.096 & 0 & 0.096 & 0 & 0.105 & 0 \\ 0 & 0 & 0.096 & 0 & 0.080 & 0 & 0 \\ 0 & 0 & 0 & 0.080 & 0 & 0.088 & 0 \\ 0 & 0 & 0.105 & 0 & 0.088 & 0 & 0.082 \\ 0 & 0 & 0 & 0 & 0 & 0.082 & 0 \end{bmatrix}$$

The distance-sum-connectivity matrix is used for computing the weighted identification number (Szymanski et al., 1986), which has been successfully tested in QSAR (Bogdanov et al., 1987; Carter et al., 1987, 1988). Randić introduced the concept of identification numbers (Randić, 1984, 1986), while Szymanski et al. (1985, 1986) investigated its discriminatory power. They found that the identification numbers are highly discriminating indices, but they are not unique. For example, in the field of 618.050 trees representing all alkanes up to 20 carbon atoms, there are 175 pairs and 20 triplets of nonisomorphic structures with the same identification number.

## 5.3   WIENER MATRICES

The *Wiener matrix* (Todeschini and Consonni, 2000, 2009), also called the *edge-Wiener matrix* (Devillers and Balaban, 1999) and denoted by $^e\mathbf{W}$, was introduced for acyclic graphs (Randić, 1993). It is a sparse symmetric square $V \times V$ matrix whose elements are defined as

$$\left[ ^e\mathbf{W} \right]_{ij} = \begin{cases} V_{i,e} \ V_{j,e} & \text{if } i \neq j \\ 0 & \text{otherwise} \end{cases} \tag{5.5}$$

where $V_{i,e}$ and $V_{j,e}$ denote the number of vertices in the two subgraphs (fragments) after an edge $i$-$j$, denoted by $e$, is removed from the acyclic graph. Below we give the edge-Wiener matrix for the branched tree $T_2$ (see Figure 2.19):

$$^e\mathbf{W}(T_2) = \begin{bmatrix} 0 & 7 & 0 & 0 & 0 & 0 & 0 & 0 \\ 7 & 0 & 15 & 0 & 0 & 0 & 0 & 7 \\ 0 & 15 & 0 & 15 & 0 & 0 & 7 & 0 \\ 0 & 0 & 15 & 0 & 12 & 0 & 0 & 0 \\ 0 & 0 & 0 & 12 & 0 & 7 & 0 & 0 \\ 0 & 0 & 0 & 0 & 7 & 0 & 0 & 0 \\ 0 & 0 & 7 & 0 & 0 & 0 & 0 & 0 \\ 0 & 7 & 0 & 0 & 0 & 0 & 0 & 0 \end{bmatrix}$$

Summation of nonzero elements in the upper or lower matrix-triangles gives the Wiener index (Wiener number). The edge-Wiener matrix was also found to be a rich and stimulating source of novel molecular descriptors (Randić et al., 1993, 1994; Lučić et al., 2003).

The sparse Wiener matrix can be made dense by considering paths instead of edges when building up the matrix elements (Randić, 1993). This type of the Wiener matrix is called the *path-Wiener matrix* (Devillers and Balaban, 1999; Todeschini and Consonni, 2009) and is denoted by $^p\mathbf{W}$. Its elements for a tree are defined as

$$\left[\,^p\mathbf{W}\right]_{ij} = \begin{cases} V_{i,p}\ V_{j,p} & \text{if } i \neq j \\ 0 & \text{otherwise} \end{cases} \tag{5.6}$$

where $p$ denotes a path between vertices $i$ and $j$. Here $V_{i,p}$ and $V_{j,p}$ represent the number of vertices on each side of path $p$, including vertices $i$ and $j$. The path-Wiener matrix for the branched tree $T_2$ (see Figure 2.19) is given as follows:

$$^p\mathbf{W}(T_2) = \begin{bmatrix} 0 & 7 & 5 & 3 & 2 & 1 & 1 & 1 \\ 7 & 0 & 15 & 9 & 6 & 3 & 3 & 7 \\ 5 & 15 & 0 & 15 & 10 & 5 & 7 & 5 \\ 3 & 9 & 15 & 0 & 12 & 6 & 3 & 3 \\ 2 & 6 & 10 & 12 & 0 & 7 & 2 & 2 \\ 1 & 3 & 5 & 6 & 7 & 0 & 1 & 1 \\ 1 & 3 & 7 & 3 & 2 & 1 & 0 & 1 \\ 1 & 7 & 5 & 3 & 2 & 1 & 1 & 0 \end{bmatrix}$$

Summation of elements in the upper or lower matrix-triangle gives the hyper-Wiener number (Randić, 1993; Randić et al., 1993, 1994; Ivanciuc and Ivanciuc, 1999; Xu and Trinajstić, 2011).

Diudea (1996a, 1996b) has also defined the *Wiener difference matrix*, denoted by $^d\mathbf{W}$, as follows:

$$^d\mathbf{W} = \,^P\mathbf{W} - \,^e\mathbf{W} \tag{5.7}$$

The nonzero elements of this matrix represent counts of paths larger than unity. The Wiener difference matrix for $T_2$ (see Figure 2.19) is exemplified below:

$$^d\mathbf{W}(T_2) = \begin{bmatrix} 0 & 0 & 5 & 3 & 2 & 1 & 1 & 1 \\ 0 & 0 & 0 & 9 & 6 & 3 & 3 & 0 \\ 5 & 0 & 0 & 0 & 10 & 5 & 0 & 5 \\ 3 & 9 & 0 & 0 & 0 & 6 & 3 & 3 \\ 2 & 6 & 10 & 0 & 0 & 0 & 2 & 2 \\ 1 & 3 & 5 & 6 & 0 & 0 & 1 & 1 \\ 1 & 3 & 0 & 3 & 2 & 1 & 0 & 1 \\ 1 & 0 & 5 & 3 & 2 & 1 & 1 & 0 \end{bmatrix}$$

Diudea and Gutman (1998) reviewed a family of the Wiener-type indices and their use in the structure-property modeling. The molecular descriptors they discussed, besides the Wiener index, are the Cluj, Harary, Kirchhoff, Schultz, and Szeged indices. All of them are mentioned in the present book and the corresponding matrices reviewed.

## 5.4  THE MODIFIED WIENER MATRICES

The *modified edge-Wiener matrix*, denoted by $^{me}\mathbf{W}$, is defined as (Nikolić et al., 2001b)

$$\left[ ^{me}\mathbf{W} \right]_{ij} = \begin{cases} 1/(V_{i,e} \ V_{j,e}) & \text{if } i \neq j \\ 0 & \text{otherwise} \end{cases} \tag{5.8}$$

Consequently, the modified edge-Wiener matrix $^{me}\mathbf{W}$ can be directly obtained from the edge-Wiener matrix $^e\mathbf{W}$ for the nonvanishing matrix elements:

$$[^{me}\mathbf{W}]_{ij} = 1/[^e\mathbf{W}]_{ij} \tag{5.9}$$

As an example, the modified Wiener matrix for the branched tree $T_2$ (see Figure 2.19) is given below:

$$^{me}\mathbf{W}(T_2) = \begin{bmatrix} 0 & 1/7 & 0 & 0 & 0 & 0 & 0 & 0 \\ 1/7 & 0 & 1/15 & 0 & 0 & 0 & 0 & 1/7 \\ 0 & 1/15 & 0 & 1/15 & 0 & 0 & 1/7 & 0 \\ 0 & 0 & 1/15 & 0 & 1/12 & 0 & 0 & 0 \\ 0 & 0 & 0 & 1/12 & 0 & 1/7 & 0 & 0 \\ 0 & 0 & 0 & 0 & 1/7 & 0 & 0 & 0 \\ 0 & 0 & 1/7 & 0 & 0 & 0 & 0 & 0 \\ 0 & 1/7 & 0 & 0 & 0 & 0 & 0 & 0 \end{bmatrix}$$

Summation of nonzero elements in the upper or lower matrix-triangle gives the modified edge-Wiener index.

The *modified path-Wiener matrix* is given by

$$\left[ ^{mp}\mathbf{W} \right]_{ij} = \begin{cases} 1/V_{i,p} \ V_{j,p} & \text{if } i \neq j \\ 0 & \text{otherwise} \end{cases} \tag{5.10}$$

$$[^{mp}\mathbf{W}]_{ij} = 1/[^p\mathbf{W}]_{ij} \tag{5.11}$$

As an example, we give below the modified path-Wiener matrix of the branched tree $T_2$ (see Figure 2.19):

$$
^{mp}\mathbf{W}(T_2) =
\begin{bmatrix}
0 & 1/7 & 1/5 & 1/3 & 1/2 & 1 & 1 & 1 \\
1/7 & 0 & 1/15 & 1/9 & 1/6 & 1/3 & 1/3 & 1/7 \\
1/5 & 1/15 & 0 & 1/15 & 1/10 & 1/5 & 1/7 & 1/5 \\
1/3 & 1/9 & 1/15 & 0 & 1/12 & 1/6 & 1/3 & 1/3 \\
1/2 & 1/6 & 1/10 & 1/12 & 0 & 1/7 & 1/2 & 1/2 \\
1 & 1/3 & 1/5 & 1/6 & 1/7 & 0 & 1 & 1 \\
1 & 1/3 & 1/7 & 1/3 & 1/2 & 1 & 0 & 1 \\
1 & 1/7 & 1/5 & 1/3 & 1/2 & 1 & 1 & 0
\end{bmatrix}
$$

A generalization of the modified Wiener matrix leading to a class of modified Wiener indices has also been proposed (Gutman et al., 2004).

## 5.5   THE REVERSE-WIENER MATRIX

The *reverse-Wiener matrix*, denoted by **RW**, is a symmetric $V \times V$ matrix, defined by means of the vertex-distance matrix $^v\mathbf{D}$ (Randić, 1997a; Balaban et al., 2000):

$$
[\mathbf{RW}]_{ij} =
\begin{cases}
D - [^v\mathbf{D}]_{ij} & \text{if } i \neq j \\
0 & \text{otherwise}
\end{cases}
\tag{5.12}
$$

where $D$ is the diameter of a graph. The diameter of a graph $G$ is the longest geodesic distance between any two vertices $i$ and $j$ in $G$, i.e., the largest $[^v\mathbf{D}]_{ij}$ value in the vertex-distance matrix (Harary, 1971). As an example of the reverse-Wiener matrix, the **RW** matrix for $T_2$ (see Figure 2.19) is given below:

$$
\mathbf{RW}(T_2) =
\begin{bmatrix}
0 & 4 & 3 & 2 & 1 & 0 & 2 & 3 \\
4 & 0 & 4 & 3 & 2 & 1 & 3 & 4 \\
3 & 4 & 0 & 4 & 3 & 2 & 4 & 3 \\
2 & 3 & 4 & 0 & 4 & 3 & 3 & 2 \\
1 & 2 & 3 & 4 & 0 & 4 & 2 & 1 \\
0 & 1 & 2 & 3 & 4 & 0 & 1 & 0 \\
2 & 3 & 4 & 3 & 2 & 1 & 0 & 2 \\
3 & 4 & 3 & 2 & 1 & 0 & 2 & 0
\end{bmatrix}
$$

Molecular descriptors such as the reverse-Wiener index and the reverse-distance sum can be obtained from the reverse-Wiener matrix. They were used to derive the structure-property models for the following alkane properties (Balaban et al., 2000): boiling points at normal pressure, molar heat capacity at 300 K, standard Gibbs energy of formation in the gas phase at 300 K, vaporization enthalpy at 300 K, refractive index at 298 K, and density at 298 K.

## 5.6  THE REVERSE-DETOUR MATRIX

In parallel to the reverse-Wiener matrix, the *reverse-detour matrix*, denoted by $^R\mathbf{DM}$, can be defined by means of the vertex-detour matrix:

$$\left[\,^R\mathbf{DM}\,\right]_{ij} = \begin{cases} LE - [\mathbf{DM}]_{ij} & \text{if } i \neq j \\ 0 & \text{otherwise} \end{cases} \tag{5.13}$$

where $LE$ is the longest elongation (detour distance) in a graph. As an example of the reverse-vertex-detour matrix, $^R\mathbf{DM}$ matrix for $G_1$ (see structure $A$ in Figure 2.1) is given below:

$$^R\mathbf{DM}(G_1) = \begin{bmatrix} 0 & 5 & 4 & 1 & 2 & 1 & 0 \\ 5 & 0 & 5 & 2 & 3 & 2 & 1 \\ 4 & 5 & 0 & 3 & 4 & 3 & 2 \\ 1 & 2 & 3 & 0 & 3 & 4 & 3 \\ 2 & 3 & 4 & 3 & 0 & 3 & 2 \\ 1 & 2 & 3 & 4 & 3 & 0 & 5 \\ 0 & 1 & 2 & 3 & 2 & 5 & 0 \end{bmatrix}$$

Molecular descriptors such as the reverse-detour index can be obtained from the reverse-detour matrix.

## 5.7  SZEGED MATRICES

Several Szeged matrices have been proposed in the literature (Diudea et al., 1997). Here, we will consider the edge Szeged matrix, the path Szeged matrix, and the Szeged difference matrix.

The *edge-Szeged matrix*, denoted by $^e\mathbf{SZ}$, was introduced by Gutman (1994a). It is formally defined as the Wiener matrix:

$$\left[\,^e\mathbf{SZ}\,\right]_{ij} = \begin{cases} V_{i,e}\ V_{j,e} & \text{if } i \neq j \\ 0 & \text{otherwise} \end{cases} \tag{5.14}$$

where $V_{i,e}$ and $V_{j,e}$ now denote the numbers of vertices lying *closer* to $i$ and to $j$, respectively, while the vertices *equidistant* to $i$ and $j$ are not counted. A consequence of the formal identity between (5.5) and (5.14) is that the edge-Wiener matrices and the edge-Szeged matrices are identical for acyclic graphs. Therefore, the edge-Szeged matrix may be regarded as the extension of the edge-Wiener matrix to cycle-containing graphs. To illustrate this, we give below the $^e\mathbf{SZ}$ matrix for $G_1$ (see structure $A$ in Figure 2.1):

$$^e\mathbf{SZ}(G_1) = \begin{bmatrix} 0 & 6 & 0 & 0 & 0 & 0 & 0 \\ 6 & 0 & 10 & 0 & 0 & 0 & 0 \\ 0 & 10 & 0 & 10 & 0 & 12 & 0 \\ 0 & 0 & 10 & 0 & 12 & 0 & 0 \\ 0 & 0 & 0 & 12 & 0 & 10 & 0 \\ 0 & 0 & 12 & 0 & 10 & 0 & 6 \\ 0 & 0 & 0 & 0 & 0 & 6 & 0 \end{bmatrix}$$

The summation of elements in the upper or lower matrix-triangle produces the edge-Szeged index (Khadikar et al., 1995; Gutman and Klavžar, 1995; Žerovnik, 1999). Žerovnik (1996) devised an algorithm for computing the edge-Szeged index of an arbitrary graph. Similarly, Dobrynin and Gutman (1996) developed an efficient method for computing the Szeged index of unbranched condensed benzenoids. Zhou and his coworkers also considered a novel version of the edge-Szeged index (Dong et al., 2011). Khadikar and his coworkers (Khadikar et al., 2005) reviewed various uses of the Szeged index for modeling physical and chemical properties, as well as for modeling physiological and pharmacological activities of organic compounds

Diudea and coworkers (Diudea et al., 1997) introduced the *path-Szeged matrix*, denoted by $^p\mathbf{SZ}$. The above definition for the edge-Szeged matrix is extended to the path-Szeged matrix by consideration of the paths instead of the edges in $G$. Now the $^p\mathbf{SZ}$ matrix is *no longer* identical to the $^p\mathbf{W}$ matrix, as can be seen by comparing these two matrices for $T_2$ (see Figure 2.19). The path-Szeged matrix for $T_2$ is given below:

$$^p\mathbf{SZ}(T_2) = \begin{bmatrix} 0 & 7 & 5 & 15 & 9 & 15 & 15 & 1 \\ 7 & 0 & 15 & 9 & 15 & 10 & 3 & 7 \\ 5 & 15 & 0 & 15 & 10 & 12 & 7 & 5 \\ 15 & 9 & 15 & 0 & 12 & 6 & 3 & 15 \\ 9 & 15 & 10 & 12 & 0 & 7 & 15 & 9 \\ 15 & 10 & 12 & 6 & 7 & 0 & 10 & 15 \\ 15 & 3 & 7 & 3 & 15 & 10 & 0 & 15 \\ 1 & 7 & 5 & 15 & 9 & 15 & 15 & 0 \end{bmatrix}$$

We also give the path-Szeged matrix for $G_1$ (see structure $A$ in Figure 2.1):

$$^{p}\mathbf{SZ}(G_1) = \begin{bmatrix} 0 & 6 & 5 & 10 & 8 & 10 & 6 \\ 6 & 0 & 10 & 4 & 12 & 6 & 12 \\ 5 & 10 & 0 & 10 & 3 & 12 & 4 \\ 10 & 4 & 10 & 0 & 12 & 2 & 10 \\ 8 & 12 & 3 & 12 & 0 & 10 & 2 \\ 10 & 6 & 12 & 2 & 10 & 0 & 6 \\ 6 & 12 & 4 & 10 & 2 & 6 & 0 \end{bmatrix}$$

The summation of elements in the upper or lower matrix-triangle produces the hyper-Szeged index (Todeschini and Consonni, 2000, 2009).

The *Szeged difference matrix*, denoted by $^{d}\mathbf{S}$, is defined as the difference of the edge- and path-Szeged matrices:

$$^{d}\mathbf{SZ} = {}^{p}\mathbf{SZ} - {}^{e}\mathbf{SZ} \tag{5.15}$$

The Szeged difference matrices for $T_2$ (see Figure 2.19) and $G_1$ (see structure $A$ in Figure 2.1) are exemplified below:

$$^{d}\mathbf{SZ}(T_2) = \begin{bmatrix} 0 & 0 & 5 & 15 & 9 & 15 & 15 & 1 \\ 0 & 0 & 0 & 9 & 15 & 10 & 3 & 0 \\ 5 & 0 & 0 & 0 & 10 & 12 & 0 & 5 \\ 15 & 9 & 0 & 0 & 0 & 6 & 3 & 15 \\ 9 & 15 & 10 & 0 & 0 & 0 & 15 & 9 \\ 15 & 10 & 12 & 6 & 0 & 0 & 10 & 15 \\ 15 & 3 & 0 & 3 & 15 & 10 & 0 & 1 \\ 1 & 0 & 5 & 15 & 9 & 15 & 15 & 0 \end{bmatrix}$$

$$^{d}\mathbf{SZ}(G_1) = \begin{bmatrix} 0 & 0 & 5 & 10 & 8 & 10 & 6 \\ 0 & 0 & 0 & 4 & 12 & 6 & 12 \\ 5 & 0 & 0 & 0 & 3 & 0 & 4 \\ 10 & 4 & 0 & 0 & 0 & 2 & 10 \\ 8 & 12 & 3 & 0 & 0 & 0 & 2 \\ 10 & 6 & 0 & 2 & 0 & 0 & 0 \\ 6 & 12 & 4 & 10 & 2 & 0 & 0 \end{bmatrix}$$

Research on the Szeged index and its variants was nicely summarized by Todeschini and Consonni (2000, 2009) in their handbooks and in the review article by Khadikar and his coworkers (Khadikar et al., 2005),

## 5.8  RECIPROCAL SZEGED MATRICES

The *reciprocal Szeged matrices*, denoted by $\mathbf{SZ}^{-1}$, are matrices whose off-diagonal elements are the reciprocal of the corresponding elements of the Szeged matrices, discussed above:

$$[\mathbf{SZ}^{-1}]_{ij} = [\mathbf{SZ}]_{ij}^{-1} \tag{5.16}$$

All elements equal to zero in the Szeged matrices are left unchanged in the reciprocal Szeged matrices. Below we give the reciprocal Szeged matrices of the edge-Szeged matrix, the path-Szeged matrix, and the Szeged difference matrix of $G_1$ (see structure $A$ in Figure 2.1):

$$
{}^{e}\mathbf{SZ}^{-1}(G_1) =
\begin{bmatrix}
0 & 1/6 & 0 & 0 & 0 & 0 & 0 \\
1/6 & 0 & 1/10 & 0 & 0 & 0 & 0 \\
0 & 1/10 & 0 & 1/10 & 0 & 1/12 & 0 \\
0 & 0 & 1/10 & 0 & 1/12 & 0 & 0 \\
0 & 0 & 0 & 1/12 & 0 & 1/10 & 0 \\
0 & 0 & 1/12 & 0 & 1/10 & 0 & 1/6 \\
0 & 0 & 0 & 0 & 0 & 1/6 & 0
\end{bmatrix}
$$

$$
{}^{p}\mathbf{SZ}^{-1}(G_1) =
\begin{bmatrix}
0 & 1/6 & 1/5 & 1/10 & 1/8 & 1/10 & 1/6 \\
1/6 & 0 & 1/10 & 1/4 & 1/12 & 1/6 & 1/12 \\
1/5 & 1/10 & 0 & 1/10 & 1/3 & 1/12 & 1/4 \\
1/10 & 1/4 & 1/10 & 0 & 1/12 & 1/2 & 1/10 \\
1/8 & 1/12 & 1/3 & 1/12 & 0 & 1/10 & 1/2 \\
1/10 & 1/6 & 1/12 & 1/2 & 1/10 & 0 & 1/6 \\
1/6 & 1/12 & 1/4 & 1/10 & 1/2 & 1/6 & 0
\end{bmatrix}
$$

$$
{}^{d}\mathbf{SZ}^{-1}(G_1) =
\begin{bmatrix}
0 & 0 & 1/5 & 1/10 & 1/8 & 1/10 & 1/6 \\
0 & 0 & 0 & 1/4 & 1/12 & 1/6 & 1/12 \\
1/5 & 0 & 0 & 0 & 1/3 & 0 & 1/4 \\
1/10 & 1/4 & 0 & 0 & 0 & 1/2 & 1/10 \\
1/8 & 1/12 & 1/3 & 0 & 0 & 0 & 1/2 \\
1/10 & 1/6 & 0 & 1/2 & 0 & 0 & 0 \\
1/6 & 1/12 & 1/4 & 1/10 & 1/2 & 0 & 0
\end{bmatrix}
$$

These matrices have not been used so far to generate molecular descriptors.

## 5.9 THE UNSYMMETRIC SZEGED MATRIX

The *unsymmetric Szeged matrix*, denoted $^u\mathbf{SZ}$, is an unsymmetric $V \times V$ matrix defined as

$$\left[^u\mathbf{SZ}\right]_{ij} = \begin{cases} V_{i,(ij)} & \text{if } i \neq j \\ 0 & \text{otherwise} \end{cases} \tag{5.17}$$

where $V_{i,(ij)}$ is the number of vertices lying closer to the vertex $i$ than to the vertex $j$ (Khadikar et al., 1995, Diudea, 1997a, 1997b). Below we give the unsymmetric Szeged matrices for $G_1$ (see structure $A$ in Figure 2.1) and $T_2$ (see Figure 2.19):

$$^u\mathbf{SZ}(G_1) = \begin{bmatrix} 0 & 1 & 1 & 2 & 2 & 2 & 2 \\ 6 & 0 & 2 & 2 & 3 & 2 & 4 \\ 5 & 5 & 0 & 5 & 3 & 4 & 4 \\ 5 & 2 & 2 & 0 & 4 & 1 & 5 \\ 4 & 4 & 1 & 3 & 0 & 2 & 2 \\ 5 & 3 & 3 & 2 & 5 & 0 & 6 \\ 3 & 3 & 1 & 2 & 1 & 1 & 0 \end{bmatrix}$$

$$^u\mathbf{SZ}(T_2) = \begin{bmatrix} 0 & 1 & 1 & 3 & 3 & 5 & 3 & 1 \\ 7 & 0 & 3 & 3 & 5 & 5 & 3 & 7 \\ 5 & 5 & 0 & 5 & 5 & 6 & 7 & 5 \\ 5 & 3 & 3 & 0 & 6 & 6 & 3 & 5 \\ 3 & 3 & 2 & 2 & 0 & 7 & 3 & 3 \\ 3 & 2 & 2 & 1 & 1 & 0 & 2 & 3 \\ 5 & 1 & 1 & 1 & 5 & 5 & 0 & 5 \\ 1 & 1 & 1 & 3 & 3 & 5 & 3 & 0 \end{bmatrix}$$

The use of unsymmetric Szeged matrices is discussed by Todeschini and Consonni (2009).

## 5.10 CLUJ MATRICES

The *Cluj matrix* is a square unsymmetrical $V \times V$ matrix, denoted by $^u\mathbf{C}$, whose elements are defined as (Diudea, 1996a, 1997a, 1997b, 1997c, 1997d; Diudea and Gutman, 1998; Diudea et al., 1998, 2007):

$$\left[^u\mathbf{C}\right]_{ij} = \begin{cases} V_{i,p(ij)} & \text{if } i \neq j \\ 0 & \text{otherwise} \end{cases} \tag{5.18}$$

where $V_{i,p(ij)}$ is determined from the subgraph $G - [p(i, j)]$ remaining when the edges and any internal vertices of the path $p(i, j)$ are deleted from $G$. Then $V_{i,p(ij)}$ is the number of vertices closer to $i$ than $j$ in the component of $G - [p(i, j)]$ containing $i$. The unsymmetric Cluj matrix can be constructed for any connected graph. In the case of graphs containing cycles, there can exist several shortest paths. Among them, one selects that path which allows the maximum value of $V_{i,p(ij)}$.

The unsymmetrical Cluj matrix for the branched tree $T_2$ (see Figure 2.19) is given below:

$$
{}^{u}\mathbf{C}(T_2) =
\begin{bmatrix}
0 & 1 & 1 & 1 & 1 & 1 & 1 & 1 \\
7 & 0 & 3 & 3 & 3 & 3 & 3 & 7 \\
5 & 5 & 0 & 5 & 5 & 5 & 7 & 5 \\
3 & 3 & 3 & 0 & 6 & 6 & 3 & 3 \\
2 & 2 & 2 & 2 & 0 & 7 & 2 & 2 \\
1 & 1 & 1 & 1 & 1 & 0 & 1 & 1 \\
1 & 1 & 1 & 1 & 1 & 1 & 0 & 1 \\
1 & 1 & 1 & 1 & 1 & 1 & 1 & 0
\end{bmatrix}
$$

The unsymmetrical Cluj matrix for $G_1$ (see structure $A$ in Figure 2.1) is as follows:

$$
{}^{u}\mathbf{C}(G_1) =
\begin{bmatrix}
0 & 1 & 1 & 1 & 1 & 1 & 1 \\
6 & 0 & 2 & 2 & 2 & 2 & 2 \\
5 & 5 & 0 & 5 & 3 & 4 & 4 \\
4 & 2 & 2 & 0 & 4 & 1 & 4 \\
3 & 3 & 1 & 3 & 0 & 2 & 2 \\
4 & 3 & 3 & 2 & 5 & 0 & 6 \\
1 & 1 & 1 & 1 & 1 & 1 & 0
\end{bmatrix}
$$

The unsymmetrical Cluj matrix ${}^{u}\mathbf{C}$ may be symmetrized to give the *path-Cluj matrix* ${}^{p}\mathbf{C}$ and the *edge-Cluj matrix* ${}^{e}\mathbf{C}$:

$$
{}^{p}\mathbf{C} = {}^{u}\mathbf{C} \cdot {}^{u}\mathbf{C}^{T} \tag{5.19}
$$

$$
{}^{e}\mathbf{C} = {}^{p}\mathbf{C} \cdot {}^{v}\mathbf{A} \tag{5.20}
$$

where the symbol $\cdot$ denotes the *Hadamard matrix product* (Horn and Johnson, 1985) and ${}^{u}\mathbf{C}^{T}$ is the transpose of ${}^{u}\mathbf{C}$.

The Hadamard matrix product of two matrices $\mathbf{A}$ and $\mathbf{B}$ of the same dimensions is defined as

$$
[\mathbf{A} \cdot \mathbf{B}]_{ij} = [\mathbf{A}]_{ij}[\mathbf{B}]_{ij} \tag{5.21}
$$

It should be noted that for acyclic graphs the ${}^{p}\mathbf{C}$ and ${}^{e}\mathbf{C}$ matrices are identical to the ${}^{p}\mathbf{W}$ and ${}^{e}\mathbf{W}$ matrices, respectively.

For cycle-containing graphs, the $^eC$ matrix is equal to the edge-Szeged matrix (Ivanciuc and Ivanciuc, 1999):

$$^eC = {}^eSZ \tag{5.22}$$

while the $^pC$ matrix differs from any previously known special matrix based on paths.
Below we give the path-Cluj matrix of $G_1$:

$$^pC(G_1) = \begin{bmatrix}
0 & 6 & 5 & 4 & 3 & 4 & 1 \\
6 & 0 & 10 & 4 & 6 & 6 & 2 \\
5 & 10 & 0 & 10 & 3 & 12 & 4 \\
4 & 4 & 10 & 0 & 12 & 2 & 4 \\
3 & 6 & 3 & 12 & 0 & 10 & 2 \\
4 & 6 & 12 & 2 & 10 & 0 & 6 \\
1 & 2 & 4 & 4 & 2 & 6 & 0
\end{bmatrix}$$

## 5.11   RECIPROCAL CLUJ MATRICES

The reciprocal Cluj matrices are matrices whose entries are the reciprocal of the corresponding entries of the symmetric Cluj matrices (Diudea and Gutman, 1998; Diudea et al., 1998). Here we present the *reciprocal edge-Cluj matrix*, denoted by $^eC^{-1}$:

$$[^eC^{-1}]_{ij} = [^eC]_{ij}^{-1} \tag{5.23}$$

and the *reciprocal path-Cluj matrix*, denoted by $^pC^{-1}$:

$$[^pC^{-1}]_{ij} = [^pC]_{ij}^{-1} \tag{5.24}$$

However, the reciprocal edge-Cluj matrix is equal to the reciprocal edge-Szeged matrix:

$$^eC^{-1} = {}^eSZ^{-1} \tag{5.25}$$

As an example we give the reciprocal path-Cluj matrix of $G_1$ (see structure $A$ in Figure 2.1):

$$^pC^{-1}(G_1) = \begin{bmatrix}
0 & 1/6 & 1/5 & 1/4 & 1/3 & 1/4 & 1 \\
1/6 & 0 & 1/10 & 1/4 & 1/6 & 1/6 & 1/2 \\
1/5 & 1/10 & 0 & 1/10 & 1/3 & 1/12 & 1/4 \\
1/4 & 1/4 & 1/10 & 0 & 1/12 & 1/2 & 1/4 \\
1/3 & 1/6 & 1/3 & 1/12 & 0 & 1/10 & 1/2 \\
1/4 & 1/6 & 1/12 & 1/2 & 1/10 & 0 & 1/6 \\
1 & 1/2 & 1/4 & 1/4 & 1/2 & 1/6 & 0
\end{bmatrix}$$

## 5.12   THE HOSOYA MATRIX

The Hosoya matrix was introduced by Randić (1994a). Denoted by $\mathbf{Z}$, the Hosoya matrix is derived here in a manner similar to the edge-Wiener matrix (see Section 5.3). It is a *sparse* symmetric square $V \times V$ matrix whose elements for a tree are defined as

$$[\mathbf{Z}]_{ij} = \begin{cases} Z_i Z_j & \text{if vertices } i \text{ and } j \text{ are adjacent} \\ 0 & \text{otherwise} \end{cases} \tag{5.26}$$

where $Z_i$ and $Z_j$ are the Hosoya Z-indices (Hosoya, 1971) of the two subgraphs (fragments) after an edge $i\text{-}j$ is removed from the acyclic graph. The Z-indices are tabulated for $C_1$ to $C_9$ fragments (Trinajstić et al., 1991), and programs for computing this and many other molecular descriptors are available (e.g., Ivanciuc and Devillers, 1999). Below we give the Hosoya matrix for the branched tree $T_2$ (see Figure 2.19):

$$\mathbf{Z}(T_2) = \begin{bmatrix} 0 & 19 & 0 & 0 & 0 & 0 & 0 & 0 \\ 19 & 0 & 24 & 0 & 0 & 0 & 0 & 19 \\ 0 & 24 & 0 & 21 & 0 & 0 & 18 & 0 \\ 0 & 0 & 21 & 0 & 20 & 0 & 0 & 0 \\ 0 & 0 & 0 & 20 & 0 & 17 & 0 & 0 \\ 0 & 0 & 0 & 0 & 17 & 0 & 0 & 0 \\ 0 & 0 & 18 & 0 & 0 & 0 & 0 & 0 \\ 0 & 19 & 0 & 0 & 0 & 0 & 0 & 0 \end{bmatrix}$$

The Hosoya matrix may be made dense if the elements $[\mathbf{Z}]_{ij}$ are computed not only for deleted edges, but also for deleted edges along any path in a tree (Randić, 1994a). Then, the *dense Hosoya matrix* $^d\mathbf{Z}$ is defined as

$$\left[ {}^d\mathbf{Z} \right]_{ij} = \begin{cases} Z_i Z_j & \text{if } i \neq j \\ 0 & \text{otherwise} \end{cases} \tag{5.27}$$

The dense Hosoya matrix for $T_2$ (see Figure 2.19) is exemplified below:

$$^d\mathbf{Z}(T_2) = \begin{bmatrix} 0 & 19 & 16 & 12 & 8 & 4 & 10 & 11 \\ 19 & 0 & 24 & 18 & 12 & 6 & 15 & 19 \\ 16 & 24 & 0 & 21 & 14 & 7 & 18 & 16 \\ 12 & 18 & 21 & 0 & 20 & 10 & 12 & 12 \\ 8 & 12 & 14 & 20 & 0 & 17 & 8 & 4 \\ 4 & 6 & 7 & 10 & 17 & 0 & 4 & 4 \\ 10 & 15 & 18 & 12 & 8 & 4 & 0 & 10 \\ 11 & 19 & 16 & 12 & 4 & 4 & 10 & 0 \end{bmatrix}$$

The Hosoya matrices are used to produce a variety of molecular descriptors, especially since the Z-index and the Hosoya matrix have been extended to poly-cyclic systems and edge-weighted graphs (e.g., Nikolić et al., 1992; Plavšić et al., 1997; Espeso et al., 2000; Miličević et al., 2003). Randić (1994a) tested success-fully the Z-indices in the structure-boiling point modeling of octanes. Similarly, Hosoya (2002) used his index for predicting the octane numbers of heptanes and octanes. Mathematical properties of the Hosoya Z-indices have also been studied (e.g., Vukičević and Trinajstić, 2005).

Hosoya elaborated the potentiality of his descriptor in his talk at the International Symposium on Applications of Mathematical Concepts in Chemistry (Dubrovnik, Croatia, September 2–5, 1985) and in his article in the book entitled *Mathematics and Computational Concepts in Chemistry* (Hosoya, 1986). Hosoya also produced in 2012 a monograph on his index (Hosoya, 2012), published in Japanese. Perhaps it will be translated into English.

## 5.13   THE PATH MATRIX

The *path matrix*, denoted by **P**, has been introduced by Randić (1991a; Randić and Trinajstić, 1993) and used by Randić for structural ordering and branching of acyclic saturated hydrocarbons (Randić, 1997a, 1997b, 1998).

The **P** matrix of a vertex-labeled connected simple graph $G$ is a square $V \times V$ matrix whose entries are defined as follows:

$$\left[ {}^{p}\mathbf{P} \right]_{ij} = \begin{cases} p'(i,j)/p & \text{if } i \neq j \\ 0 & \text{otherwise} \end{cases} \tag{5.28}$$

where $p'(i, j)$ is the total number of paths in the subgraph $G'$ obtained by remov-ing the edge $i$-$j$ from $G$, and $p$ is the total number of paths in $G$. Randić suggested that if the subgraph $G'$ is disjoint, then the contribution of each component should be added. However, a few years earlier, Mekenyan et al. (1988) derived topological indices for molecular fragments and multiplied the contribution by each component. In presenting examples of the path matrix, we follow the Randić suggestion, that is, the addition of contributions. In Figure 5.1, we give the $p'(i, j)/p$ values of $T_2$ (see Figure 2.19) and $G_1$ (see structure $A$ in Figure 2.1).

The corresponding path matrices are exhibited below:

$$\mathbf{P}(T_2) = \begin{bmatrix} 0 & 21/28 & 0 & 0 & 0 & 0 & 0 & 0 \\ 21/28 & 0 & 13/28 & 0 & 0 & 0 & 0 & 21/28 \\ 0 & 13/28 & 0 & 13/28 & 0 & 0 & 21/28 & 0 \\ 0 & 0 & 13/28 & 0 & 15/28 & 0 & 0 & 0 \\ 0 & 0 & 0 & 15/28 & 0 & 21/28 & 0 & 0 \\ 0 & 0 & 0 & 0 & 21/28 & 0 & 0 & 0 \\ 0 & 0 & 21/28 & 0 & 0 & 0 & 0 & 0 \\ 0 & 21/28 & 0 & 0 & 0 & 0 & 0 & 0 \end{bmatrix}$$

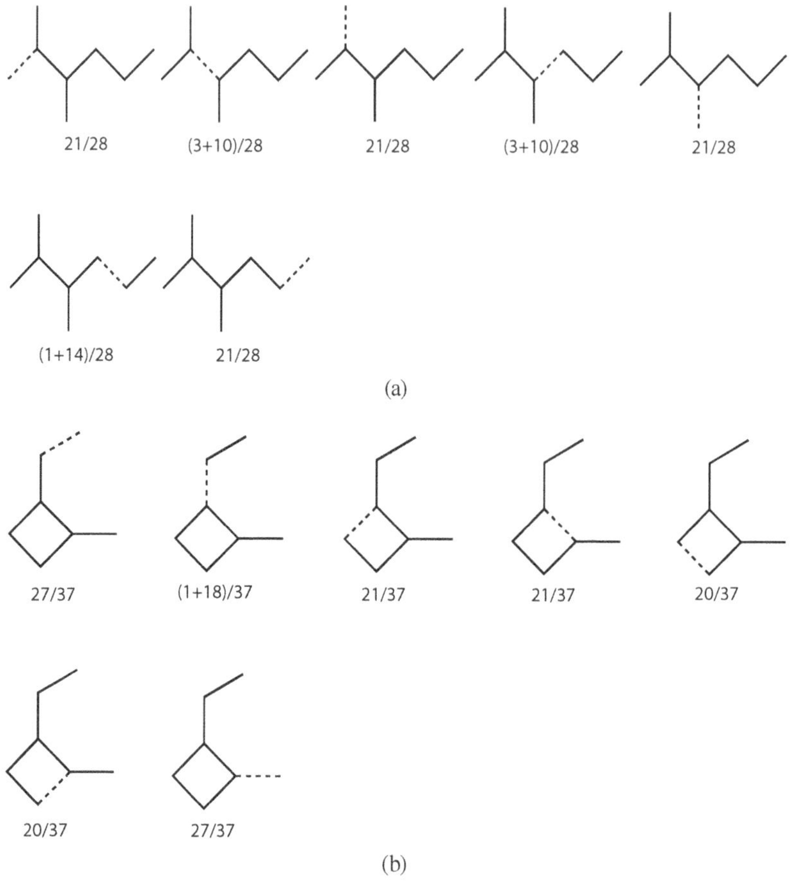

21/28        (3+10)/28        21/28        (3+10)/28        21/28

(1+14)/28        21/28

(a)

27/37        (1+18)/37        21/37        21/37        20/37

20/37        27/37

(b)

**FIGURE 5.1** Entries to the path matrices of $T_2$ (a) and $G_1$ (b). The broken lines represent the removed edges.

$$P(G_1) = \begin{bmatrix} 0 & 27/37 & 0 & 0 & 0 & 0 & 0 \\ 27/37 & 0 & 19/37 & 0 & 0 & 0 & 0 \\ 0 & 19/37 & 0 & 21/37 & 0 & 21/37 & 0 \\ 0 & 0 & 21/37 & 0 & 20/37 & 0 & 0 \\ 0 & 0 & 0 & 20/37 & 0 & 20/37 & 0 \\ 0 & 0 & 21/37 & 0 & 20/37 & 0 & 27/37 \\ 0 & 0 & 0 & 0 & 0 & 27/37 & 0 \end{bmatrix}$$

The quantity $p'(i, j)/p$ could be considered a graphical bond order of the edge $i$-$j$ in $G$ (Randić et al., 1997). By summing up the nonvanishing entries in the upper (or the lower) matrix-triangle, the $P'/P$ index is obtained (Randić, 1991b). It has also been shown that the $P'/P$ index and the Wiener index are closely related graph-theoretical invariants of trees and cycles (Plavšić et al., 1996).

## 5.14   THE ALL-PATH MATRIX

The *all-path matrix*, denoted by $^\wedge \mathbf{P}$, is a square symmetric $V \times V$ matrix whose $i\text{-}j$ entry is the sum of the lengths of all paths $p(i, j)$ connecting vertices $i$ and $j$ (Todeschini and Consonni, 2000, 2009):

$$\left[ ^\wedge \mathbf{P} \right]_{ij} = \begin{cases} \sum |p(i,j)| & \text{if } i \neq j \\ 0 & \text{otherwise} \end{cases} \tag{5.29}$$

where $|p(i, j)|$ denotes the length of a path between vertices $i$ and $j$. The all-path matrix is a graph-theoretical representation of molecules based on the total path-count.

From the above definition, it follows that for simple acyclic graphs the all-path matrix is identical to the vertex-distance matrix. For acyclic graphs, the total path-count $P$ is given by a simple expression:

$$P = \frac{V^2 - V}{2} \tag{5.30}$$

The all-path matrix of $G_1$ (see structure $A$ in Figure 2.1) is exemplified below:

$$^\wedge \mathbf{P}(G_1) = \begin{bmatrix} 0 & 1 & 2 & 8 & 8 & 8 & 10 \\ 1 & 0 & 1 & 6 & 6 & 6 & 8 \\ 2 & 1 & 0 & 4 & 4 & 4 & 6 \\ 8 & 6 & 4 & 0 & 4 & 4 & 6 \\ 8 & 6 & 4 & 4 & 0 & 4 & 6 \\ 8 & 6 & 4 & 4 & 4 & 0 & 1 \\ 10 & 8 & 6 & 6 & 6 & 1 & 0 \end{bmatrix}$$

The all-path matrix is used for the computation of the all-path Wiener index (Lukovits, 1998).

## 5.15   THE EXPANDED VERTEX-DISTANCE MATRICES

The *expanded vertex-distance matrix*, denoted by $^v \mathbf{Д}$, has been introduced by Tratch et al. (1990). It is a square symmetric $V \times V$ matrix defined as (Ivanciuc et al., 1997)

$$\left[ ^v \mathbf{Д} \right]_{ij} = \begin{cases} N_{ij} \, [\mathbf{D}]_{ij} & \text{if } i \neq j \\ 0 & \text{otherwise} \end{cases} \tag{5.31}$$

where $N_{ij}$ is the number of all shortest paths containing the path $p(i, j)$ as a subpath. The matrix $^v Д$ for $G_1$ (see structure $A$ in Figure 2.1) is given by

$$
^v Д(G_1) = \begin{bmatrix}
0 & 7 & 12 & 6 & 4 & 9 & 4 \\
7 & 0 & 12 & 8 & 12 & 12 & 6 \\
12 & 12 & 0 & 6 & 12 & 9 & 6 \\
6 & 8 & 6 & 0 & 4 & 8 & 6 \\
4 & 12 & 12 & 4 & 0 & 5 & 4 \\
9 & 12 & 9 & 8 & 5 & 0 & 7 \\
4 & 6 & 6 & 6 & 4 & 7 & 0
\end{bmatrix}
$$

For acyclic graphs, the expanded vertex-distance matrix is given by

$$
[^v Д]_{ij} = \begin{cases} l(i, j) \cdot V_{i,p} \cdot V_{j,p} & \text{if } i \neq j \\ 0 & \text{otherwise} \end{cases} \tag{5.32}
$$

Or in other words, the elements of the matrix $^v Д$ for acyclic graphs can be simply obtained from the elements of $^v \mathbf{D}$ and $^p \mathbf{W}$ by the Hadamard matrix product:

$$
^v Д = {}^v \mathbf{D} \cdot {}^p \mathbf{W} \tag{5.33}
$$

The expanded vertex-distance matrix of $T_2$ (see Figure 2.19) is as follows:

$$
^v Д(T_2) = \begin{bmatrix}
0 & 7 & 10 & 9 & 8 & 5 & 3 & 2 \\
7 & 0 & 15 & 18 & 18 & 12 & 6 & 7 \\
10 & 15 & 0 & 15 & 20 & 15 & 7 & 10 \\
9 & 18 & 15 & 0 & 12 & 12 & 6 & 9 \\
8 & 18 & 20 & 12 & 0 & 7 & 6 & 8 \\
5 & 12 & 15 & 12 & 7 & 0 & 4 & 5 \\
3 & 6 & 7 & 6 & 6 & 4 & 0 & 3 \\
2 & 7 & 10 & 9 & 8 & 5 & 3 & 0
\end{bmatrix}
$$

If the path-Wiener matrix, $^p \mathbf{W}$, is substituted by the unsymmetric Cluj matrix, then the unsymmetric expanded vertex-distance matrix is obtained. This has been used for a new definition of the hyper-Wiener index (Diudea and Gutman, 1998; Ivanciuc et al., 1998; Gutman, 2004). This kind of expanded vertex-distance matrix of $T_2$ is shown below:

$$
{}^{v}\textit{Д}(T_2) = \begin{bmatrix}
0 & 1 & 2 & 3 & 4 & 5 & 3 & 2 \\
7 & 0 & 3 & 6 & 9 & 12 & 6 & 7 \\
10 & 5 & 0 & 5 & 10 & 15 & 7 & 10 \\
9 & 6 & 3 & 0 & 6 & 12 & 6 & 9 \\
8 & 6 & 4 & 2 & 0 & 7 & 6 & 8 \\
5 & 4 & 3 & 2 & 1 & 0 & 4 & 5 \\
3 & 2 & 1 & 2 & 3 & 4 & 0 & 3 \\
2 & 1 & 2 & 3 & 4 & 5 & 3 & 0
\end{bmatrix}
$$

Diudea and Gutman (1998) generalized the concept of the expanded vertex-distance matrix in order to define novel matrices using the Hadamard matrix product between the vertex-distance matrix ${}^{v}\mathbf{D}$ and a general square $V \times V$ matrix $\mathbf{\Psi}$ as

$$
{}^{v}\textit{Д}\,\mathbf{\Psi} = {}^{v}\mathbf{D} \cdot \mathbf{\Psi} \tag{5.34}
$$

If $\mathbf{\Psi}$ is a Szeged matrix, for example, the expanded vertex-distance Szeged matrix is derived. Todeschini and Consonni in their *Handbook of Molecular Descriptor* (Todeschini and Consonni, 2000, Table E-8) and in *Molecular Descriptors for Chemoinformatics* (Todeschini and Consonni, 2009, Table E-12) listed 28 expanded vertex-distance matrices. From these matrices, two kinds of molecular descriptors can be derived, i.e., expanded distance indices and expanded square distance indices.

## 5.16   THE QUOTIENT MATRICES

Quotient matrices have been introduced by Randić (1994b) and applied to structure-property modeling by Plavšić and his coworkers (Plavšić et al., 1998) and Nikolić and her coworkers (Nikolić et al., 2001a). The *quotient matrices*, denoted by $\mathbf{M}_a/\mathbf{M}_b$, are obtained by dividing the off-diagonal elements of matrices $\mathbf{M}_a$ and $\mathbf{M}_b$:

$$
\left[\mathbf{M}_a/\mathbf{M}_b\right]_{ij} = \begin{cases} [\mathbf{M}_a]_{ij}/[\mathbf{M}_b]_{ij} & \text{if } i \neq j \\ 0 & \text{otherwise} \end{cases} \tag{5.35}
$$

Several quotient matrices are in use. Here we list six: the vertex-distance/detour matrix ${}^{v}\mathbf{D}/\mathbf{DM}$ (Randić, 1994b), the detour/vertex-distance matrix $\mathbf{DM}/{}^{v}\mathbf{D}$ (Plavšić et al., 1998), the vertex-distance/resistance-distance matrix ${}^{v}\mathbf{D}/\Omega$ (Babić et al., 2002; Klein and Ivanciuc, 2002), the resistance-distance/vertex-distance matrix $\Omega/{}^{v}\mathbf{D}$ (Babić et al., 2002; Klein and Ivanciuc, 2002), the vertex-distance/vertex-distance-complement matrix ${}^{v}\mathbf{D}/{}^{vc}\mathbf{D}$ (Nikolić et al., 2001a), and the vertex-distance-complement/vertex-distance matrix ${}^{vc}\mathbf{D}/{}^{v}\mathbf{D}$ (Nikolić et al., 2001a). These six quotient matrices for $G_1$ (see structure A in Figure 2.1) are given below:

$$
{}^{v}\mathbf{D/DM}(G_1) = \begin{bmatrix}
0 & 1 & 1 & 0.6 & 1 & 0.6 & 0.67 \\
1 & 0 & 1 & 0.5 & 1 & 0.5 & 0.6 \\
1 & 1 & 0 & 0.33 & 1 & 0.33 & 0.5 \\
0.6 & 0.5 & 0.33 & 0 & 0.33 & 1 & 1 \\
1 & 1 & 1 & 0.33 & 0 & 0.33 & 0.5 \\
0.6 & 0.5 & 0.33 & 1 & 0.33 & 0 & 1 \\
0.67 & 0.6 & 0.5 & 1 & 0.5 & 1 & 0
\end{bmatrix}
$$

The molecular index based on the vertex-distance/detour matrix is called the Wiener-sum index (Randić, 1994b):

$$
\mathbf{DM/}{}^{v}\mathbf{D}(G_1) = \begin{bmatrix}
0 & 1 & 1 & 1.67 & 1 & 1.67 & 1.5 \\
1 & 0 & 1 & 2 & 1 & 2 & 1.67 \\
1 & 1 & 0 & 3 & 1 & 3 & 2 \\
1.67 & 2 & 3 & 0 & 3 & 1 & 1 \\
1 & 1 & 1 & 3 & 0 & 3 & 2 \\
1.67 & 2 & 3 & 1 & 3 & 0 & 1 \\
1.5 & 1.67 & 2 & 1 & 2 & 1 & 0
\end{bmatrix}
$$

The molecular index based on the detour/vertex-distance matrix is called the detour-sum index (Plavšić et al., 1998):

$$
{}^{v}\mathbf{D/\Omega}(G_1) = \begin{bmatrix}
0 & 1 & 1 & 1.09 & 1.33 & 1.09 & 1.07 \\
1 & 0 & 1 & 1.14 & 1.5 & 1.14 & 1.09 \\
1 & 1 & 0 & 1.33 & 2 & 1.33 & 1.14 \\
1.09 & 1.14 & 1.33 & 0 & 1.33 & 2 & 1.5 \\
1.33 & 1.5 & 2 & 1.33 & 0 & 1.33 & 1.14 \\
1.09 & 1.14 & 1.33 & 2 & 1.33 & 0 & 1 \\
1.07 & 1.09 & 1.14 & 1.5 & 1.14 & 1 & 0
\end{bmatrix}
$$

The molecular index based on the vertex-distance/resistance-distance matrix is a variant of the Wiener-sum index (Babić et al., 2002):

$$
\mathbf{\Omega/}{}^{v}\mathbf{D}(G_1) = \begin{bmatrix}
0 & 1 & 1 & 0.92 & 0.75 & 0.92 & 0.93 \\
1 & 0 & 1 & 0.88 & 0.67 & 0.88 & 0.92 \\
1 & 1 & 0 & 0.75 & 0.5 & 0.75 & 0.88 \\
0.92 & 0.88 & 0.75 & 0 & 0.75 & 0.5 & 0.67 \\
0.75 & 0.67 & 0.5 & 0.75 & 0 & 0.75 & 0.88 \\
0.92 & 0.88 & 0.75 & 0.5 & 0.75 & 0 & 1 \\
0.93 & 0.92 & 0.88 & 0.67 & 0.88 & 1 & 0
\end{bmatrix}
$$

The molecular index based on the resistance-distance/vertex-distance matrix is called the Kirchhoff-sum index (Babić et al., 2002). Matrices $^v\mathbf{D}/\Omega$ and $\Omega/^v\mathbf{D}$ have been used to study the graph cyclicity (Klein and Ivanciuc, 2002):

$$
^v\mathbf{D}/^{vc}\mathbf{D}(G_1) = \begin{bmatrix}
0 & 0.2 & 0.5 & 1 & 2 & 1 & 2 \\
0.2 & 0 & 0.2 & 0.5 & 1 & 0.5 & 1 \\
0.5 & 0.2 & 0 & 0.2 & 0.5 & 0.2 & 0.5 \\
1 & 0.5 & 0.2 & 0 & 0.2 & 0.5 & 1 \\
2 & 1 & 0.5 & 0.2 & 0 & 0.2 & 0.5 \\
1 & 0.5 & 0.2 & 0.5 & 0.2 & 0 & 0.2 \\
2 & 1 & 0.5 & 1 & 0.5 & 0.2 & 0
\end{bmatrix}
$$

$$
^{vc}\mathbf{D}/^v\mathbf{D}(G_1) = \begin{bmatrix}
0 & 5 & 2 & 1 & 0.5 & 1 & 0.5 \\
5 & 0 & 5 & 2 & 1 & 2 & 1 \\
2 & 5 & 0 & 5 & 2 & 5 & 2 \\
1 & 2 & 5 & 0 & 5 & 2 & 1 \\
0.5 & 1 & 2 & 5 & 0 & 5 & 2 \\
1 & 2 & 5 & 2 & 5 & 0 & 5 \\
0.5 & 1 & 2 & 1 & 2 & 5 & 0
\end{bmatrix}
$$

Two variants of the Balaban index (Balaban, 1989) have been derived from the $^v\mathbf{D}/^{vc}\mathbf{D}$ and $^{vc}\mathbf{D}/^v\mathbf{D}$ matrices (Nikolić et al., 2001a).

## 5.17   THE RANDOM-WALK MARKOV MATRIX

A *random walk* in a (molecular) graph is naturally designated by a probability measure, of which there are at least two natural intrinsic examples. The first is the probability measure that entails starting walks from each vertex with equal probability and taking subsequent steps such that each neighboring vertex is reached with equal probability, so that the probability of stepping from one vertex to another vertex is $1/d$, where $d$ is the vertex-degree. The walks of the consequently generated distribution are referred to (Doyle and Snell, 1984) as *simple random* walks. The second is the probability measure that takes each possible walk of a given length as equally probable. These probability measures are generally quite different, though if a graph is regular, i.e., having all vertices of the same degree, then the two probability measures are equivalent. The walks generated by this second probability measure are called *equipoise random* walks (Klein et al., 2004). The simple random walks have rarely been used in chemical graph theory (e.g., Shalabi, 1991; Palacios, 2001), whereas the equipoise random walks have been used more often (e.g., Marcus, 1963; Randić, 1980; Randić et al., 1983; Todeschini and Consonni, 2000, 2009; Rücker and Rücker, 1991, 1993, 1999, 2000, 2001, 2003; Gutman et al., 2001; Lukovits et al., 2002; Lukovits and Trinajstić, 2003; Nikolić et al., 2003b). Often either equipoise or simple random walks have been called just *random*, probably without recognition of the alternative type—or perhaps as a result of confusion of the two possibilities.

Walks can be generated from powers of the vertex-adjacency matrix $^vA$ (see Section 2.1), and this may be viewed as an identification of the distribution for equipoise random walks. Similarly, the distribution for simple random walks can be generated by powers of a Markov matrix. The *random-walk Markov matrix*, denoted by **MM**, of a vertex-labeled connected graph $G$ is a real unsymmetrical $V \times V$ matrix whose elements are probabilities for the associated individual steps (Klein et al., 2004):

$$[\mathbf{MM}]_{ij} = \begin{cases} 1/[d(j)] & \text{if } i \neq j \\ 0 & \text{otherwise} \end{cases} \tag{5.36}$$

Then $[\mathbf{MM}^\lambda]_{ij}$ is the probability for a $\lambda$-step random walk beginning at vertex $j$ and ending at vertex $i$.

The random-walk Markov matrices of the vertex-labeled $T_2$ (see Figure 2.19) and graph $G_1$ (see structure $A$ in Figure 2.1) is as follows:

$$\mathbf{MM}(T_2) = \begin{bmatrix} 0 & 1/3 & 0 & 0 & 0 & 0 & 0 & 0 \\ 1 & 0 & 1/3 & 0 & 0 & 0 & 0 & 1 \\ 0 & 1/3 & 0 & 1/2 & 0 & 0 & 1 & 0 \\ 0 & 0 & 1/3 & 0 & 1/2 & 0 & 0 & 0 \\ 0 & 0 & 0 & 1/2 & 0 & 1 & 0 & 0 \\ 0 & 0 & 0 & 0 & 1/2 & 0 & 0 & 0 \\ 0 & 0 & 1/3 & 0 & 0 & 0 & 0 & 0 \\ 0 & 1/3 & 0 & 0 & 0 & 0 & 0 & 0 \end{bmatrix}$$

$$\mathbf{MM}(G_1) = \begin{bmatrix} 0 & 1/2 & 0 & 0 & 0 & 0 & 0 \\ 1 & 0 & 1/3 & 0 & 0 & 0 & 0 \\ 0 & 1/2 & 0 & 1/2 & 0 & 1/3 & 0 \\ 0 & 0 & 1/3 & 0 & 1/2 & 0 & 0 \\ 0 & 0 & 0 & 1/2 & 0 & 1/3 & 0 \\ 0 & 0 & 1/3 & 0 & 1/2 & 0 & 1 \\ 0 & 0 & 0 & 0 & 0 & 1/3 & 0 \end{bmatrix}$$

The Markov matrix **MM** may be also expressed in terms of the vertex-adjacency matrix $^vA$ and the inverse diagonal matrix $\mathbf{\Delta}$:

$$\mathbf{MM} = {}^v\mathbf{A}\,\mathbf{\Delta}^{-1} \tag{5.37}$$

Several unsymmetrical graph-theoretical matrices such as the Cluj matrices (see Section 5.10) and the layer matrices (Diudea and Ursu, 2003) have also been proposed. However, the Markov matrix has a much wider field of application (e.g., Ross, 1977). It should be noted that the (simple) random walks have been extensively studied in mathematics and physics (Doyle and Snell, 1984), but only occasionally in chemistry, as has already been noted.

Algebraic manipulations with the Markov matrix appear to be rewarding in chemical graph theory. For example, the combination of the Markov matrix **MM** and the diagonal matrix $\mathbf{\Delta}^{1/2}$ with elements

$$[\mathbf{\Delta}^{1/2}]_{ii} = [d(i)]^{1/2} \tag{5.38}$$

and its inverse $\mathbf{\Delta}^{-1/2}$, gives the following symmetric matrix:

$$\mathbf{\Gamma} = \mathbf{\Delta}^{-1/2}\, \mathbf{MM}\, \mathbf{\Delta}^{1/2} \tag{5.39}$$

As an example, we give below this matrix for a vertex-labeled graph $G_1$ (see structure A in Figure 2.1):

$$\mathbf{\Gamma}(G_1) = \begin{bmatrix}
0 & 1/\sqrt{2} & 0 & 0 & 0 & 0 & 0 \\
1/\sqrt{2} & 0 & 1/\sqrt{6} & 0 & 0 & 0 & 0 \\
0 & 1/\sqrt{6} & 0 & 1/\sqrt{6} & 0 & 0 & 0 \\
0 & 0 & 1/\sqrt{6} & 0 & 1/2 & 0 & 0 \\
0 & 0 & 0 & 1/2 & 0 & 1/\sqrt{6} & 0 \\
0 & 0 & 0 & 0 & 1/\sqrt{6} & 0 & 1/\sqrt{3} \\
0 & 0 & 0 & 0 & 0 & 1/\sqrt{3} & 0
\end{bmatrix}$$

Matrix $\mathbf{\Gamma}$ is interesting for at least two reasons (Klein et al., 2004). The first reason is that the half-sum of elements of the $\mathbf{\Gamma}$ matrix gives the connectivity index $\chi$ (Randić, 1975):

$$\chi = (1/2)\sum_{i \ne j} (\mathbf{\Gamma})_{ij} \tag{5.40}$$

thus allowing the interpretation of this molecular descriptor as a sum of symmetrized neighbor-hopping probabilities. The second is that it leads to the (standard or combinatorial) Laplacian matrix **L**:

$$\mathbf{L} = \mathbf{\Delta}^{1/2}\, (\mathbf{I} - \mathbf{\Gamma})\, \mathbf{\Delta}^{1/2} \tag{5.41}$$

while the matrix $\mathbf{I} - \boldsymbol{\Gamma}$ is sometimes called the analytic or normalized Laplacian matrix (Chung, 1997). If we denote this normalized Laplacian matrix by $\mathbf{L}_{norm}$, the connectivity index $\chi$ can also be expressed in terms of its elements:

$$\chi = (1/2)\sum_{i \neq j}(L_{norm})_{ij}$$

(5.42)

## 5.18   RESTRICTED RANDOM-WALK MATRIX

The *restricted random-walk matrix*, denoted by **RRW**, has been introduced by Randić (1995). The **RRW** matrix is defined as

$$[\mathbf{RRW}]_{ij} = \begin{cases} 1/P_{i \to j} & \text{if } i \neq j \\ 0 & \text{otherwise} \end{cases}$$

(5.43)

where $P_{i \to j}$ is the probability that a random walk leaving site $i$ will reach site $j$ before returning to $i$.

The number of steps in a random walk is restricted by the distance between the vertices considered. Let's consider restricted random walks on the tree $T_2$ (see Figure 2.19). For example, if one begins a walk at vertex 1 and wishes to arrive at vertex 3, which is only two edges away, random walks are restricted to length 2. Thus, the following restricted random walks of length 2 are allowed: 1-2-3, 1-2-8, and 1-2-1. There are three such walks, but only one walk is successful; thus, the probability to reach vertex 3 from vertex 1 is 1/3. However, if we consider walks from vertex 3 to vertex 1, the number of possible walks of length 2 starting at vertex 3 increases: 3-2-1, 3-2-8, 3-2-3, 3-4-5, 3-4-3, and 3-7-3. The probability of reaching vertex 1 from vertex 3 is 1/6. As the consequence of this, the matrix **RRW** is asymmetric.

The **RRW** matrix for the tree $T_2$ (see Figure 2.19) is given below:

$$\mathbf{RRW}(T_2) = \begin{bmatrix} 0 & 1 & 1/3 & 1/5 & 1/12 & 1/23 & 1/3 & 1/5 \\ 1/3 & 0 & 1/3 & 1/5 & 1/12 & 1/23 & 1/3 & 1/5 \\ 1/6 & 1/3 & 0 & 1/3 & 1/6 & 1/13 & 1/6 & 1/3 \\ 1/9 & 1/6 & 1/2 & 0 & 1/2 & 1/6 & 1/9 & 1/6 \\ 1/12 & 1/7 & 1/3 & 1/2 & 0 & 1/2 & 1/12 & 1/7 \\ 1/12 & 1/7 & 1/3 & 1/4 & 1 & 0 & 1/12 & 1/7 \\ 1/6 & 1/3 & 1 & 1/3 & 1/6 & 1/13 & 0 & 1/6 \\ 1/3 & 1 & 1/3 & 1/5 & 1/12 & 1/23 & 1/5 & 0 \end{bmatrix}$$

Randić (1995) used selected invariants (paths of different lengths) of the restricted random-walk matrices for successfully deriving the structure-entropy model of octanes.

## 5.19   THE TRANSFER MATRIX

The *transfer matrix*, denoted by **T**, is a very useful matrix for computing the number of Kekulé structures (1-factors, dimers) of long strips with repeating units (unit cells) such as polymeric chains or cage structures such as fullerenes that are built up by fusing several symmetry-equivalent units (Lieb, 1967; Trinajstić, 1983, 1992; Klein et al., 1986a, 1986b, 1986c, 2002; Babić and Graovac, 1986; Trinajstić et al., 1986; Seitz and Schmalz, 1990; Cvetković et al., 1995). A diagram of a polyphenanthrene strip is given in Figure 5.2.

The elements of the transfer matrix represent the propagation of a Kekulé structure from a position at one side of a unit cell to the other side of the cell. When the different possible local features of the Kekulé structures at the boundary of a cell are indicated, the local structure at each boundary is specified. In Figure 5.2, the local structures are at each of the positions marked by a transverse broken line. At these positions there are only two types of local structures possible, labeled by $a$ and $b$. The final step to establish the structure of the transfer matrix involves the determination of the number of ways to propagate from one local structure to another at the unit-cell boundary.

If we consider the polyphenanthrene strip given in Figure 5.2, it is seen that at a position immediately following a local structure $a$, there can occur either two $a$ local structures (because the number of Kekulé structures for the benzene ring separating two $a$ local structures is 2) or a single $b$ local structure. In the case of a local structure $b$, there can at the position immediately following it occur either $a$ or $b$ local structures. This can be summarized as

$$a \Rightarrow 2a + b \tag{5.44}$$

$$b \Rightarrow a + b \tag{5.45}$$

If $K_L(a)$ or $K_L(b)$ denotes the number of Kekulé structures for a polyphenanthrene strip with the length L and with local structures $a$ or $b$ at the terminal end, then

$$K_{L+1}(a) = 2\,K_L(a) + K_L(b) \tag{5.46}$$

$$K_{L+1}(b) = K_L(a) + K_L(b) \tag{5.47}$$

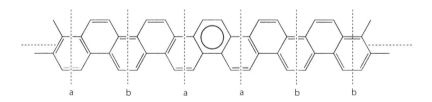

**FIGURE 5.2**   A part of the polyphenanthrene strip and the corresponding local structures.

and after setting up the transfer matrix $\mathbf{T}$,

$$\mathbf{T} = \begin{bmatrix} 2 & 1 \\ 1 & 1 \end{bmatrix}$$

it follows that

$$\begin{bmatrix} K_{L+1}(a) \\ K_{L+1}(b) \end{bmatrix} = \mathbf{T} \begin{bmatrix} K_L(a) \\ K_L(b) \end{bmatrix} = \mathbf{T}^L \begin{bmatrix} K_1(a) \\ K_1(b) \end{bmatrix} \tag{5.48}$$

The final counting formula is given by

$$K = \begin{bmatrix} 2 & 1 \end{bmatrix} \mathbf{T}^L \begin{bmatrix} 1 \\ 1 \end{bmatrix} \tag{5.49}$$

Then, the standard matrix multiplication produces the number of Kekulé structures. For example, the number of Kekulé structures for the polyphenanthrene strip of length 6 is 987.

The transfer matrix (Klein et al., 1990; Klein, 2002; Lukovits and Janežič, 2004) was also used to enumerate the conjugated circuits (Randić, 1976, 1977a, 1977b, 2003; Klein et al., 1987) of conjugated polymers, fullerenes, nanotubes, and even graphite. The transfer matrix method can be used further to enumerate a great variety of subgraph structures, especially for regularly repeating polymeric or quasi-one-dimensional parent graphs. As a number of researchers have shown, this includes matchings, self-avoiding walks, characteristic polynomials (i.e., weighted Sachs subgraphs (Mallion et al., 1974a, 1974b, 1975; Graovac et al., 1975; Trinajstić, 1983, 1992)), resonance-theoretic subgraphs, and rather general Ising-type-model subgraphs. Even further, much the same idea applies (Tesár and Fillo, 1988) in treating coupled differential equations on suitable quasi-one-dimensional systems.

## REFERENCES

D. Babić and A. Graovac, Enumeration of Kekulé structures in one-dimensional polymers, *Croat. Chem. Acta* 59 (1986) 731.744.

D. Babić, D.J. Klein, I. Lukovits, S. Nikolić, and N. Trinajstić, Resistance-distance matrix: A computational algorithm and its application, *Int. J. Quantum Chem.* 90 (2002) 166–176.

A.T. Balaban, Highly discriminating distance-based topological index, *Chem. Phys. Lett.* 89 (1989) 399–404.

A.T. Balaban, D. Mills, O. Ivanciuc, and S.C. Basak, Reverse Wiener indices, *Croat. Chem. Acta* 73 (2000) 923–941.

B. Bogdanov, S. Nikolić, A. Sabljić, N. Trinajstić, and S. Carter, On the use of the weighted identification numbers in QSAR study of the toxicity of aliphatic ethers, *Int. J. Quantum Chem. Quantum Biol. Symp.* 14 (1987) 325–330.

S. Carter, N. Trinajstić, and S. Nikolić, A note on the use of ID numbers in QSAR studies, *Acta Pharm. Jugosl.* 37 (1987) 37–42.

S. Carter, N. Trinajstić, and S. Nikolić, On the use of ID numbers in drug research: A QSAR of neuroleptic pharmacophores, *Med. Sci. Res.* 16 (1988) 185–186.

F.R.K. Chung, *Spectral graph theory*, AMM, Providence, RI, 1997.

D. Cvetković, M. Doob, and H. Sachs, *Spectra of graphs—Theory and applications*, 3rd ed., Johann Ambrosius Barth Verlag, Heidelberg, 1995.

J. Devillers and A.T. Balaban, eds., *Topological indices and related descriptors in QSAR and QSPR*, Gordon & Breach, Amsterdam, 1999.

M.V. Diudea, Walk numbers $^eW_M$: Wiener-type numbers of higher rank, *J. Chem. Inf. Comput. Sci.* 36 (1996a) 535–540.

M.V. Diudea, Wiener and hyper-Wiener numbers in a single matrix, *J. Chem. Inf. Comput. Sci.* 36 (1996b) 833–836.

M.V. Diudea, Indices of reciprocal properties or Harary indices, *J. Chem. Inf. Comput. Sci.* 37 (1997a) 292–299.

M.V. Diudea, Cluj matrix invariants, *J. Chem. Inf. Comput. Sci.* 37 (1997b) 300–305.

M.V. Diudea, Cluj matrix: Source of various graph descriptors, *MATCH Commun. Math. Comput. Chem.* 35 (1997c) 169–183.

M.V. Diudea, Cluj matrix $CJ_u$: Source of various descriptors, *MATCH Commun. Math. Comput. Chem.* 35 (1997d) 169–183.

M.V. Diudea and I. Gutman, Wiener-type topological indices, *Croat. Chem. Acta* 71 (1998) 21–51.

M.V. Diudea, I. Gutman, and J. Lorentz, *Molecular topology*, Nova, Huntington, NY, 2001.

M.V. Diudea, G. Katona, I. Lukovits, and N. Trinajstić, Detour and Cluj-detour indices, *Croat. Chem. Acta* 71 (1998) 459–471.

M.V. Diudea, O.M. Minailiuc, G. Katona, and I. Gutman, Szeged matrices and related numbers, *MATCH Commun. Math. Comput. Chem.* 35 (1997) 129–143.

M.V. Diudea and O. Ursu, Layer matrices and distance property descriptors, *Ind. J. Chem.* 41A (2003) 1283–1294.

M.V. Diudea, A.E. Vizitiu, and D. Janežič, Cluj and related polynomials applied in correlating studies, *J. Chem. Inf. Model.* 47 (2007) 864–874.

A.A. Dobrynin and I. Gutman, On the Szeged index of unbranched catacondensed benzenoid molecules, *Croat. Chem. Acta* 69 (1996) 845–856.

H. Dong, B. Zhou, and N. Trinajstić, A novel version of the edge Szeged index, *Croat. Chem. Acta* 84 (2011) 543–545.

P.G. Doyle and J.L. Snell, *Random walks and electric networks*, MAA, Washington, DC, 1984.

V.G. Espeso, J. Javier, M. Vara, B. Roy Lázaro, F. Riera Parcerisas, and D. Plavšić, On the Hosoya hyperindex and the molecular indices based on a new decomposition of the Hosoya Z matrix, *Croat. Chem. Acta* 73 (2000) 1017–1026.

E. Estrada and I. Gutman, A topological index based on distances of edges of molecular graphs, *J. Chem. Inf. Comput. Sci.* 36 (1996) 850–853.

A. Graovac, O.E. Polansky, N. Trinajstić, and N. Tyutyulkov, Graph theory in chemistry. II. Graph-theoretical description of heteroconjugated molecules, *Z. Naturforsch.* 30a (1975) 1696–1699.

I. Gutman, A formula for the Wiener number of trees and its extension to graph-containing cycles, *Graph Theory Notes NY* 27 (1994a) 9–15.

I. Gutman, Selected properties of the Schultz molecular topological index, *J. Chem. Inf. Comput. Sci.* 34 (1994b) 1087–1089.

I. Gutman, A new hyper-Wiener index, *Croat. Chem. Acta* 77 (2004) 61–64.

I. Gutman and S. Klavžar, An algorithm for the calculation of the Szeged index of benzenoid hydrocarbons, *J. Chem. Inf. Comput. Sci.* 35 (1995) 1011–1014.

I. Gutman and S. Klavžar, Bounds for the Schultz molecular topological index of benzenoid systems in terms of the Wiener index, *J. Chem. Inf. Comput. Sci.* 37 (1997) 741–744.

I. Gutman, G. Rücker, and C. Rücker, On walks in molecular graphs, *J. Chem. Inf. Comput. Sci.* 41 (2001) 739–745.

I. Gutman, D. Vukičević, and J. Žerovnik, A class of modified Wiener indices, *Croat. Chem. Acta* 77 (2004) 103–109.

F. Harary, *Graph theory*, 2nd printing, Addison-Wesley, Reading, MA, 1971.

H. Hong, S. Slavov, W. Ge, F. Qian, Z. Su, H. Fang, Y. Cheng, R. Perkins, L. Shi, and W. Tong, Mold$^2$ molecular descriptors for QSAR, in *Statistical modelling of molecular descriptors in QSAR/QSPR*, ed. M. Dehmer, K. Varmuza, and D. Bonchev, Wiley-VCH Verlag, Weinheim, 2012, pp. 65–109.

R.A. Horn and C.R. Johnson, *Matrix analysis*, University Press, Cambridge, 1985.

H. Hosoya, Topological index. A newly proposed quantity characterizing the topological nature of structural isomers of saturated hydrocarbons, *Bull. Chem. Soc. Jpn.* 44 (1971) 2332–2339.

H. Hosoya, Topological index as a common tool for quantum chemistry, statistical mechanics and graph theory, in *Mathematics and computational concepts in chemistry*, ed. N. Trinajsić, Horwood/Wiley, New York, 1986, pp. 110–123.

H. Hosoya, Chemical meaning of octane number analyzed by topological indices, *Croat. Chem. Acta* 75 (2002) 433–445.

H. Hosoya, Topological index—A new mathematics linking between the Fibonacci numbers and the Pascal triangle, Nippon hyooron Sha, Tokyo, 2012. This monograph has been reviewed by Shinsaku Fujita in *Commun. Math. Comput. Chem.* 70 (2013) 719–720.

O. Ivanciuc and J. Devillers, Algorithms and software for the computation of topological indices and structure-property models, in *Topological indices and related descriptors in QSAR and QSPR*, ed. J. Devillers and A.T. Balaban, Gordon & Breach, Amsterdam, 1999, pp. 779–804.

O. Ivanciuc, M.V. Diudea, and P. Khadikar, New topological matrices and their polynomials, *Ind. J. Chem.* 37A (1998) 574–585.

O. Ivanciuc and T. Ivanciuc, Matrices and structural descriptors computed from molecular graph distances, in *Topological indices and related descriptors in QSAR and QSPR*, ed. J. Devillers and A.T. Balaban, Gordon & Breach, Amsterdam, 1999, pp. 221–277.

O. Ivanciuc, T. Ivanciuc, and M.V. Diudea, Molecular graph matrices and derived structural descriptors, *SAR QSAR Environ. Res.* 7 (1997) 63–87.

A. Jurić, M. Gagro, S. Nikolić, and N. Trinajstić, Molecular topological index: An application in the QSAR study of toxicity of alcohols, *J. Math. Chem.* 11 (1992) 179–186.

P.V. Khadikar, N.V. Deshpande, P.P. Kale, A.A. Dobrynin, I. Gutman, and G. Dömötör, The Szeged index and an analogy with the Wiener index, *J. Chem. Inf. Comput. Sci.* 35 (1995) 547–550.

P.V. Khadikar, S. Karmarkar, V.K. Agrawal, J. Singh, A. Shrivastava, I. Lukovits, and M.V. Diudea, Szeged index—Application for drug design, *Lett. Drug Des. Disc.* 2 (2005) 563–566.

D.J. Klein, Resonating valence-bond theories for carbon π-networks and classical/quantum connections, in *Valence bond theory*, ed. D.L. Cooper, Elsevier, Amsterdam, 2002, pp. 447–502.

D.J. Klein, D. Babić, and N. Trinajstić, Enumeration in chemistry, in *Chemical modelling: Applications and theory*, Vol. 2, ed. A. Hinchliffe, Royal Society of Chemistry, Cambridge, 2002, pp. 56–95.

D.J. Klein, G.E. Hite, W.A. Seitz, and T.G. Schmalz, Dimer coverings and Kekulé structures on honeycomb lattice strips, *Theoret. Chim. Acta* 69 (1986a) 409–423.

D.J. Klein and O. Ivanciuc, Graph cyclicity, excess conductance and resistance deficit, *J. Math. Chem.* 30 (2002) 271–288.

D.J. Klein, Z. Mihalić, D. Plavšić, and N. Trinajstić, Molecular topological index: A relation with the Wiener index, *J. Chem. Inf. Comput. Sci.* 32 (1992) 304–305.

D.J. Klein, J.L. Palacios, M. Randić, and N. Trinajstić, Random walks and chemical graph theory, *J. Chem. Inf. Comput. Sci.* 44 (2004) 1521–1525.

D.J. Klein, T.G. Schmalz, and G.E. Hite, Transfer-matrix method for subgraph enumeration: Application to polypyrene fusenes, *J. Comput. Chem.* 7 (1986b) 443–456.

D.J. Klein, T.G. Schmalz, G.E. Hite, and W.A. Seitz, Resonance in $C_{60}$, buckminsterfullerene, *J. Am. Chem. Soc.* 108 (1986c) 1301–1302.

D.J. Klein, W.A. Seitz, and T.G. Schmalz, Conjugated circuit computations for conjugated hydrocarbons, in *Computational chemical graph theory*, ed. D.H. Rouvray, Nova, New York, 1990, pp. 127–147.

D.J. Klein, T.P. Živković, and N. Trinajstić, Resonance in random $\pi$-network polymers, *J. Math. Chem.* 1 (1987) 309–334.

J. von Knop, W.R. Müller, K. Szymanski, and N. Trinajstić, On the determinant of the adjacency matrix as the topological index characterizing alkanes, *J. Chem. Inf. Comput. Sci.* 31 (1991) 83–84.

E.H. Lieb, Solution of the dimer problem by the transfer matrix method, *J. Math. Phys.* 8 (1967) 2339–2341.

B. Lučić, A. Miličević, S. Nikolić, and N. Trinajstić, On variable Wiener index, *Ind. J. Chem.* 42A (2003) 1279–1282.

I. Lukovits, An all-path version of the Wiener index, *J. Chem. Inf. Comput. Sci.* 38 (1998) 125–129.

I. Lukovits and D. Janežič, Enumeration of conjugated circuits in nanotubes, *J. Chem. Inf. Comput. Sci.* 44 (2004) 410–414.

I. Lukovits, A. Miličević, S. Nikolić, and N. Trinajstić, On walk counts and complexity of general graphs, *Internet Electronic J. Mol. Des.* 1 (2002) 388–400.

I. Lukovits and N. Trinajstić, Atomic walk counts of negative order, *J. Chem. Inf. Comput. Sci.* 43 (2003) 1110–1114.

R.B. Mallion, A.J. Schwenk, and N. Trinajstić, A graphical study of heteroconjugated molecules, *Croat. Chem. Acta* 46 (1974a) 171–182.

R.B. Mallion, A.J. Schwenk, and N. Trinajstić, On the characteristic polynomial of a rooted graph, in *Recent advances in graph theory*, ed. M. Fiedler, Academia, Prague, 1975, pp. 345–350.

R.B. Mallion, N. Trinajstić, and A.J. Schwenk, Graph theory in chemistry. Generalization of Sachs' formula, *Z. Naturforsch.* 29a (1974b) 1481–1484.

R.A. Marcus, Additivity of heats of combustion, LCAO resonance energies and bond orders of conformal sets of conjugated compounds, *J. Chem. Phys.* 43 (1963) 2643–2654.

O. Mekenyan, D. Bonchev, and A.T. Balaban, Topological indices for molecular fragments and new graph invariants, *J. Math. Chem.* 2 (1988) 347–375.

Z. Mihalić, S. Nikolić, and N. Trinajstić, Comparative study of molecular descriptors derived from the distance matrix, *J. Chem. Inf. Comput. Sci.* 32 (1992a) 28–37.

Z. Mihalić and N. Trinajstić, A graph-theoretical approach to structure-property relationships, *J. Chem. Educ.* 69 (1992) 701–712.

Z. Mihalić, D. Veljan, D. Amić, S. Nikolić, D. Plavšić, and N. Trinajstić, The distance matrix in chemistry, *J. Math. Chem.* 11 (1992b) 223–258.

A. Miličević, S. Nikolić, D. Plavšić, and N. Trinajstić, On the Hosoya Z-index of general graphs, *Internet Electron. J. Mol. Des.* 2 (2003) 160–178. http://www.biochempress.com

W.R. Müller, K. Szymanski, J. von Knop, and N. Trinajstić, Molecular topological index, *J. Chem. Inf. Comput. Sci.* 30 (1990) 160–163.

S. Nikolić, G. Kovačević, A. Miličević, and N. Trinajstić, The Zagreb indices 30 years after, *Croat. Chem. Acta* 76 (2003a) 113–124.

S. Nikolić, D. Plavšić, and N. Trinajstić, On the Z-counting polynomial for weighted graphs, *J. Math. Chem.* 9 (1992) 381–387.

S. Nikolić, D. Plavšić, and N. Trinajstić, On the Balaban-like topological indices, *MATCH Commun. Math. Comput. Chem.* 44 (2001a) 361–386.

S. Nikolić, N. Trinajstić, and M. Randić, Wiener index revisited, *Chem. Phys. Lett.* 333 (2001b) 319–321.

S. Nikolić, N. Trinajstić, I.M. Tolić, G. Rücker, and C. Rücker, On molecular complexity indices, in *Complexity in chemistry—Introduction and fundamentals*, ed. D. Bonchev and D.H. Rouvray, Taylor and Francis, London, 2003b, pp. 29–76.

J.L. Palacios, Resistance distance in graphs and random walks, *Int. J. Quantum Chem.* 81 (2001) 29–33.

D. Plavšić, S. Nikolić, N. Trinajstić, and D.J. Klein, Relation between the Wiener index and the Schultz index for several classes of chemical graphs, *Croat. Chem. Acta* 66 (1993) 345–353.

D. Plavšić, M. Šoškić, Z. Đaković, I. Gutman, and A. Graovac, Extension of the Z matrix to cycle-containing and edge-weighted molecular graphs, *J. Chem. Inf. Comput. Sci.* 37 (1997) 529–534.

D. Plavšić, M. Šoškić, I. Landeka, and N. Trinajstić, On the relation between the $P'/P$ index and the Wiener number, *J. Chem. Inf. Comput. Sci.* 36 (1996) 1123–1126.

D. Plavšić, N. Trinajstić, D. Amić, and M. Šoškić, Comparison between the structure-boiling point relationships with different descriptors for condensed benzenoids, *New J. Chem.* 22 (1998) 1075–1078.

M. Randić, On characterization of molecular branching, *J. Am. Chem. Soc.* 97 (1975) 6609–6615.

M. Randić, Conjugated circuits and resonance energies of benzenoid hydrocarbons, *Chem. Phys. Lett.* 38 (1976) 68–70.

M. Randić, Aromaticity and conjugation, *J. Am. Chem. Soc.* 99 (1977a) 444–450.

M. Randić, A graph-theoretical approach to conjugation and resonance energies of hydrocarbons, *Tetrahedron* 33 (1977b) 1905–1920.

M. Randić, Random walks and their diagnostic value for characterization of atomic environment, *J. Comput. Chem.* 4 (1980) 386–399.

M. Randić, On molecular identification numbers *J. Chem. Inf. Comput. Sci.* 24 (1984) 164–175.

M. Randić, Compact molecular codes, *J. Chem. Inf. Comput. Sci.* 26 (1986) 136–148.

M. Randić, Generalized molecular descriptors, *J. Math. Chem.* 7 (1991a) 155–168.

M. Randić, On computation of optimal parameters for multivariate analysis of structure-property relationship, *J. Comput. Chem.* 12 (1991b) 970–980.

M. Randić, Novel molecular descriptor for structure-property studies, *Chem. Phys. Lett.* 211 (1993) 478–483.

M. Randić, Hosoya matrix—A source of new molecular descriptors, *Croat. Chem. Acta* 67 (1994a) 415–429.

M. Randić, On characterization of cyclic structures, *J. Chem. Inf. Comput. Sci.* 34 (1994b) 403–409.

M. Randić, Restricted random walks on graphs, *Theoret. Chim. Acta* 92 (1995) 97–106.

M. Randić, On molecular branching, *Acta Chim. Sloven.* 44 (1997a) 57–77.

M. Randić, Linear combination of path numbers as molecular descriptors, *New J. Chem.* 21 (1997b) 945–951.

M. Randić, On structural ordering and branching of acyclic saturated hydrocarbons, *J. Math. Chem.* 24 (1998) 345–358.

M. Randić, Aromaticity of polycyclic conjugated hydrocarbons, *Chem. Rev.* 103 (2003) 3449–3605.

M. Randić, X. Guo, T. Oxley, and H. Krishnapriyan, Wiener matrix: Source of novel graph invariants, *J. Chem. Inf. Comput. Sci.* 33 (1993) 709–716.

M. Randić, X. Guo, T. Oxley, H. Krishnapriyan, and L. Naylor, Wiener matrix invariants, *J. Chem. Inf. Comput. Sci.* 34 (1994) 361–367.

M. Randić, Z. Mihalić, S. Nikolić, and N. Trinajstić, Graphical bond orders: Novel structural descriptors, *J. Chem. Inf. Comput. Sci.* 37 (1997) 1063–1071.

M. Randić and N. Trinajstić, In search of graph invariants of chemical interest, *J. Mol. Struct.* 300 (1993) 551–571.

M. Randić, W.L. Woodworth, and A. Graovac, Unusual random walks, *Int. J. Quantum Chem.* 24 (1983) 435–452.

R.T. Ross, Steady state of stiff linear kinetic systems by a Markov matrix method, *J. Comput. Phys.* 24 (1977) 70–80.

G. Rücker and C. Rücker, On using the adjacency matrix power method for perception of symmetry and for isomorphism testing of highly intricate graphs, *J. Chem. Inf. Comput. Sci.* 31 (1991) 123–126.

G. Rücker and C. Rücker, Counts of all walks as atomic and molecular descriptors, *J. Chem. Inf. Comput. Sci.* 33 (1993) 683–695.

G. Rücker and C. Rücker, On topological indices, boiling points and cycloalkanes, *J. Chem. Inf. Comput. Sci.* 39 (1999) 788–802.

G. Rücker and C. Rücker, Walk counts, labyrinthicity and complexity of acyclic and cyclic graphs and molecules, *J. Chem. Inf. Comput. Sci.* 40 (2000) 99–106.

G. Rücker and C. Rücker, Substructure, subgraph and walk counts as measures of the complexity of graphs and molecules, *J. Chem. Inf. Comput. Sci.* 41 (2001) 1457–1462.

G. Rücker and C. Rücker, Walking backward: Walk counts of negative order, *J. Chem. Inf. Comput. Sci.* 43 (2003) 1115–1120.

H.P. Schultz, Topological organic chemistry. 1. Graph theory and topological indices of alkanes *J. Chem. Inf. Comput. Sci.* 29 (1989) 227–281.

H.P. Schultz, E.B. Schultz, and T.P. Schultz, Topological organic chemistry. 2. Graph theory, matrix determinants and eigenvalues, and topological indices of alkanes, *J. Chem. Inf. Comput. Sci.* 30 (1990) 27–29.

W.A. Seitz and T.G. Schmalz, Benzenoid polymers, in *Valence bond theory and chemical structure*, ed D.J. Klein and N. Trinajstić, Elsevier, Amsterdam, 1990, pp. 525–551.

P.G. Seybold, Topological influences on the carcinogenecity of aromatic hydrocarbons. I. Bay region geometry, *Int. J. Quantum Chem. Quantum Biol. Symp.* 10 (1983) 95–101.

A.S. Shalabi, Random walks: Computations and applications to chemistry, *J. Chem. Inf. Comput. Sci.* 31 (1991) 483–491.

K. Szymanski, W.R. Müller, J. von Knop, and N. Trinajstić, On Randić's molecular identification numbers, *J. Chem. Inf. Comput. Sci.* 25 (1985) 413–415.

K. Szymanski, W.R. Müller, J. von Knop, and N. Trinajstić, On the identification numbers for chemical structures, *Int. J. Quantum Chem. Quantum Chem. Symp.* 20 (1986) 173–183.

A. Tesár and L. Fillo, *Transfer matrix method*, Kluwer, Dordrecht, 1988.

R. Todeschini and V. Consonni, *Handbook of molecular descriptors*, Wiley-VCH, Weinheim, 2000.

R. Todeschini and V. Consonni, *Molecular descriptors for chemoinformatics*, Vols. I and II, Wiley-VCH, Weinheim, 2009.

S.S. Tratch, I.V. Stankevich, and N.S. Zefirov, Combinatorial models and algorithms in chemistry. The expanded Wiener number—A novel topological index, *J. Comput. Chem.* 11 (1990) 899–908.

N. Trinajstić, *Chemical graph theory*, Vols. I and II, CRC, Boca Raton, FL, 1983.

N. Trinajstić, *Chemical graph theory*, 2nd ed., CRC, Boca Raton, FL, 1992.

N. Trinajstić, D.J. Klein, and M. Randić, On some solved and unsolved problems of chemical graph theory, *Int. J. Quantum Chem. Quantum Chem. Symp.* 20 (1986) 699–742.

N. Trinajstić, S. Nikolić, J. von Knop, W.R. Müller, and K. Szymanski, *Computational chemical graph theory—Characterization, enumeration and generation of chemical structures by computer methods*, Horwood/Simon & Schuster, New York, 1991, pp. 263–266.

D. Vukičević and N. Trinajstić, Comparison of the Hosoya Z-index for simple and general graphs of the same size, *Croat. Chem. Acta* 78 (2005) 235–239.

K. Xu and N. Trinajstić, Hyper-Wiener and Harary indices of graphs with cut edges, *Utilitas Math.* 84 (2011) 153–163.

J. Žerovnik, Computing the Szeged index, *Croat. Chem. Acta* 69 (1996) 837–843.

J. Žerovnik, Szeged index of symmetric graphs, *J. Chem. Inf. Comput. Sci.* 39 (1999) 77–80.

# 6 Graphical Matrices

Graphical matrices are matrices whose elements are subgraphs of the graph rather than numbers. Since the elements of these matrices are subgraphs, they are called the *graphical matrices* (Randić et al., 2004). Thus far, limited work has been reported on these matrices (Randić et al., 1997, 2004; Nikolić et al., 2005a, 2005b; Miličević and Trinajstić, 2006; Jančžič et al., 2007). However, many of the so-called special matrices presented above, such as the Wiener matrices and the Hosoya matrices, may be regarded as the numerical realizations of the corresponding graphical matrices. The advantage of graphical matrices lies in the fact that they allow many possibilities of numerical realizations. In order to obtain a numerical form of a graphical matrix, one needs to select a graph invariant and replace all the graphical elements (subgraphs of some form) by the corresponding numerical values of the selected invariant. In this way, the numerical form of the graphical matrix is established and one can select another or the same type of invariant—this time an invariant of the numerical matrix. Graph invariants generated in this way are *double invariants* in view of the fact that two invariants are used in constructing a given molecular descriptor.

## 6.1 CONSTRUCTION OF GRAPHICAL MATRICES

Here we present two ways of constructing graphical matrices, denoted by $\mathbf{G}$, that lead to four types of these matrices. One way is to define the elements of the graphical matrix $[\mathbf{G}]_{ij}$ as the *subgraphs* obtained after the consecutive removal of *edges* connecting vertices $i$ and $j$ from the graph $G$. We denote this kind of graphical matrix by $^{e}\mathbf{G}$, where $e$ stands for the *edge*, and we call it *edge-graphical matrix*. The matrix $^{e}\mathbf{G}$ is necessarily a *sparse* matrix, since it contains only a few nonvanishing elements corresponding to the removed edges. An example of this kind of graphical matrix is given in Figure 6.1. Since the graphical matrix is a square symmetric $V \times V$ matrix, it is enough for demonstrative purposes to give only the upper triangle of the matrix. For graphs without loops, the corresponding graphical matrices have zeros as diagonal elements.

However, if we generate the graphical matrix by the consecutive removal of *paths* joining vertices $i$ and $j$ instead of edges, the obtained matrix is *dense*. We call this matrix the *path-graphical matrix* and denote it by $^{p}\mathbf{G}$. An illustrative example of a path-graphical matrix is given in Figure 6.2.

A second way to construct graphical matrices is to define their elements $[\mathbf{G}]_{ij}$ as the *subgraphs* obtained after the consecutive removal of *adjacent* vertices $i$ and $j$, and the incident edges from the graph $G$. The graphical matrices so obtained are by necessity *sparse* matrices, since they contain only a few nonvanishing elements corresponding to the deleted adjacent vertices and incident edges. We denote this kind of graphical matrix by $^{sv}\mathbf{G}$, where $s$ denotes the sparse matrix and $v$ stands for the

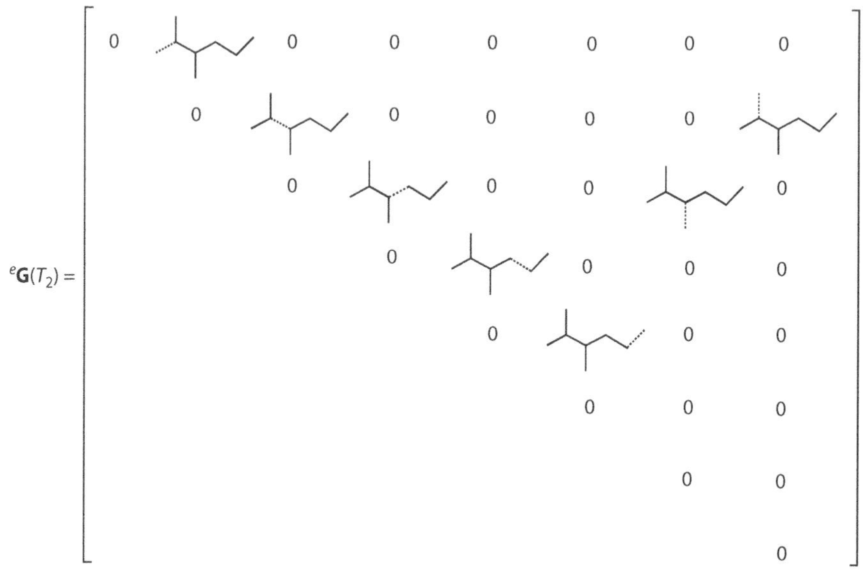

**FIGURE 6.1**   The edge-graphical matrix of $T_2$.

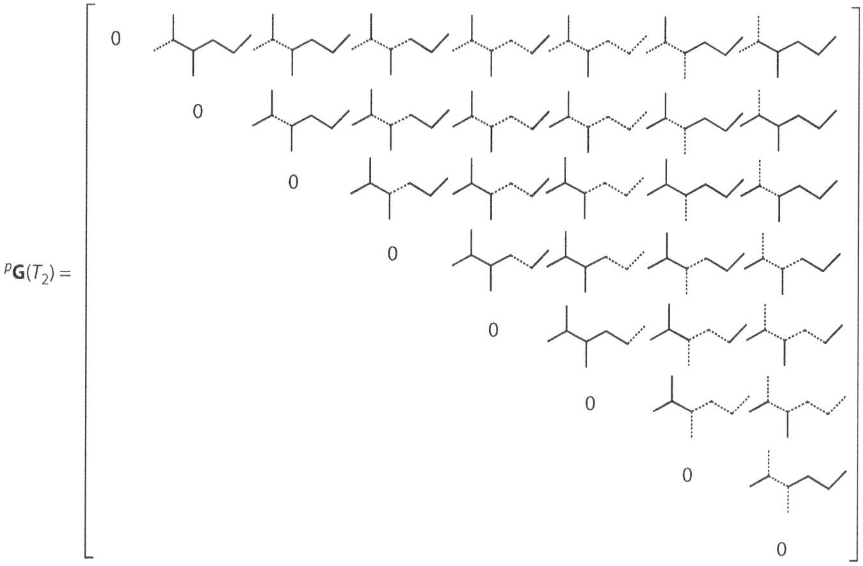

**FIGURE 6.2**   The path-graphical matrix of $T_2$.

adjacent vertices, and we call these matrices the *sparse vertex-graphical matrices*. An example of such a matrix is given in Figure 6.3.

If, instead of considering only adjacent vertices, we consider pairs of vertices $i$ and $j$ at increasing distances, the graphical matrix obtained is *dense*; that is, all its matrix-elements except the diagonal elements are nonzero. We denote this matrix by

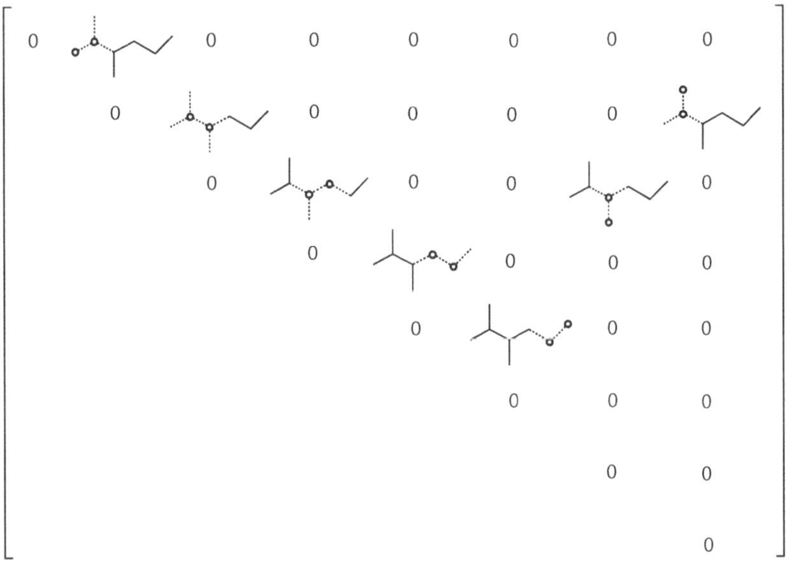

**FIGURE 6.3** The sparse vertex-graphical matrix of $T_2$.

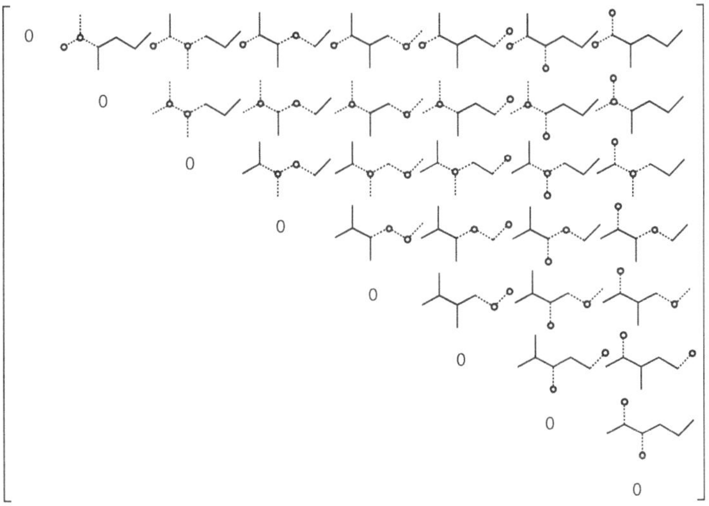

**FIGURE 6.4** The dense vertex-graphical matrix of $T_2$.

$^{dv}\mathbf{G}$, where $d$ denotes the dense matrix, and call it the *dense vertex-graphical matrix*. In Figure 6.4, we give the dense vertex-graphical matrix of $T_2$.

## 6.2 NUMERICAL REALIZATION OF GRAPHICAL MATRICES

In order to use graphical matrices, we need to replace subgraphs with the invariants of choice. To exemplify this, we employ three graph invariants often used in the structure-property-activity modeling: the *Randić connectivity index* (Randić,

1975; Li and Gutman, 2006; Gutman and Furtula, 2008), also known as the *vertex-connectivity index* and *product-connectivity index* (Trinajstić et al., 1997; Nikolić et al., 1998), the *sum-connectivity index* (Lučić et al., 2009, 2010, 2013; Zhou and Trinajstić, 2012; Gutman, 2013), and the *Hosoya index* (Hosoya, 1971). The numbers that replace the subgraphs in the graphical matrices are obtained by *summing up* (in the case of the connectivity indices) and *multiplying* (in the case of the Hosoya index) the corresponding graph invariants. There are also other ways in which to construct numerical matrices from graphical matrices. The values of the Randić connectivity indices, sum-connectivity indices, and Hosoya indices for acyclic subgraphs are taken from our book on computational chemical graph theory (Trinajstić et al., 1991, Table IX.1) and the review (Lučić et al., 2010, Table 2). Below are given three types of numerical realization of the four graphical matrices presented in Figures 6.1 to 6.4.

### 6.2.1   USE OF THE RANDIĆ CONNECTIVITY INDEX

The first numerical matrix, when the Randić connectivity index is employed, is named the *edge-Randić matrix* and denoted by $^e\mathbf{R}$. An example of this matrix, obtained from the edge-graphical matrix of $T_2$, is given below. For practical reasons, only the upper triangle of the matrix is shown.

$$
^e\mathbf{R}(T_2) =
\begin{bmatrix}
0 & 3.31 & 0 & 0 & 0 & 0 & 0 & 0 \\
 & 0 & 3.83 & 0 & 0 & 0 & 0 & 3.31 \\
 & & 0 & 3.68 & 0 & 0 & 3.27 & 0 \\
 & & & 0 & 3.64 & 0 & 0 & 0 \\
 & & & & 0 & 3.18 & 0 & 0 \\
 & & & & & 0 & 0 & 0 \\
 & & & & & & 0 & 0 \\
 & & & & & & & 0
\end{bmatrix}
$$

From this matrix can be derived, for example, the vertex-Randić-Wiener index, obtained by summing the elements of the matrix-triangle.

The second numerical matrix is named the *path-Randić matrix* and denoted by $^p\mathbf{R}$. An example of this matrix, obtained from the path-graphical matrix of $T_2$, is given below, and again, only the upper triangle of the matrix is shown:

$$
^p\mathbf{R}(T_2) =
\begin{bmatrix}
0 & 3.31 & 3.41 & 3.00 & 3.00 & 2.00 & 2.91 & 2.77 \\
 & 0 & 3.83 & 3.83 & 3.41 & 2.41 & 3.33 & 3.31 \\
 & & 0 & 3.68 & 3.27 & 2.27 & 3.27 & 3.41 \\
 & & & 0 & 3.64 & 2.64 & 3.15 & 3.41 \\
 & & & & 0 & 3.18 & 2.41 & 3.00 \\
 & & & & & 0 & 1.73 & 2.00 \\
 & & & & & & 0 & 2.91 \\
 & & & & & & & 0
\end{bmatrix}
$$

From this matrix can be obtained, for example, the path-Randić-Wiener index by summing up the elements in the matrix-triangle.

The third numerical matrix is named the *sparse vertex-Randić matrix* and denoted by $^{sv}\mathbf{R}$. An example of this matrix, obtained from the sparse vertex-graphical matrix of $T_2$, is given below. Only the upper matrix-triangle is given.

$$
^{sv}\mathbf{R}(T_2) =
\begin{bmatrix}
0 & 2.41 & 0 & 0 & 0 & 0 & 0 & 0 \\
  & 0 & 1.41 & 0 & 0 & 0 & 0 & 2.41 \\
  &   & 0 & 2.41 & 0 & 0 & 2.83 & 0 \\
  &   &   & 0 & 2.27 & 0 & 0 & 0 \\
  &   &   &   & 0 & 2.64 & 0 & 0 \\
  &   &   &   &   & 0 & 0 & 0 \\
  &   &   &   &   &   & 0 & 0 \\
  &   &   &   &   &   &   & 0
\end{bmatrix}
$$

The summation of the matrix-elements in the above matrix-triangle gives the sparse vertex-Randić-Wiener index.

The fourth numerical matrix is named the *dense vertex-Randić matrix* and denoted by $^{dv}\mathbf{R}$. An example of this matrix, obtained from the dense vertex-graphical matrix of $T_2$, is given below, and again only the upper matrix-triangle is given.

$$
^{dv}\mathbf{R}(T_2) =
\begin{bmatrix}
0 & 2.41 & 2.41 & 2.91 & 2.27 & 2.81 & 2.91 & 2.77 \\
  & 0 & 1.41 & 2.00 & 1.41 & 1.91 & 1.91 & 2.41 \\
  &   & 0 & 2.41 & 1.41 & 2.41 & 2.83 & 2.41 \\
  &   &   & 0 & 2.27 & 2.27 & 2.73 & 2.91 \\
  &   &   &   & 0 & 2.64 & 2.27 & 2.27 \\
  &   &   &   &   & 0 & 2.77 & 2.81 \\
  &   &   &   &   &   & 0 & 2.91 \\
  &   &   &   &   &   &   & 0
\end{bmatrix}
$$

The molecular descriptor based on this matrix, obtained by summing the elements in its triangle, is called the dense vertex-Randić-Wiener index.

## 6.2.2 USE OF THE SUM-CONNECTIVITY INDEX

Here we give numerically the four sum-connectivity matrices of a graph $T_2$ representing the carbon skeleton of 2,3-dimethylhexane. The first matrix is named the *edge-sum-connectivity matrix* and is denoted by $^e\mathbf{SCM}$. We give only the upper triangle of each matrix.

$$
{}^{e}\mathbf{SCM}(T_2) = \begin{bmatrix} 0 & 3.05 & 0 & 0 & 0 & 0 & 0 & 0 \\ & 0 & 3.31 & 0 & 0 & 0 & 0 & 3.05 \\ & & 0 & 3.18 & 0 & 0 & 3.03 & 0 \\ & & & 0 & 3.12 & 0 & 0 & 0 \\ & & & & 0 & 2.93 & 0 & 0 \\ & & & & & 0 & 0 & 0 \\ & & & & & & 0 & 0 \\ & & & & & & & 0 \end{bmatrix}
$$

The second matrix is named the *path-sum-connectivity matrix* and is denoted by ${}^{p}\mathbf{SCM}$:

$$
{}^{p}\mathbf{SCM}(T_2) = \begin{bmatrix} 0 & 3.05 & 2.86 & 2.83 & 2.12 & 1.41 & 2.36 & 2.53 \\ & 0 & 3.31 & 3.02 & 2.83 & 1.86 & 2.81 & 3.05 \\ & & 0 & 3.18 & 2.73 & 2.03 & 3.03 & 2.86 \\ & & & 0 & 3.12 & 2.41 & 2.66 & 2.57 \\ & & & & 0 & 2.93 & 2.21 & 2.12 \\ & & & & & 0 & 1.50 & 1.41 \\ & & & & & & 0 & 2.36 \\ & & & & & & & 0 \end{bmatrix}
$$

The third matrix is named *sparse vertex-sum-connectivity matrix* and is denoted by ${}^{sv}\mathbf{SCM}$:

$$
{}^{sv}\mathbf{SCM}(T_2) = \begin{bmatrix} 0 & 2.16 & 0 & 0 & 0 & 0 & 0 & 0 \\ & 0 & 1.16 & 0 & 0 & 0 & 0 & 2.16 \\ & & 0 & 1.86 & 0 & 0 & 2.31 & 0 \\ & & & 0 & 2.03 & 0 & 0 & 0 \\ & & & & 0 & 2.41 & 0 & 0 \\ & & & & & 0 & 0 & 0 \\ & & & & & & 0 & 0 \\ & & & & & & & 0 \end{bmatrix}
$$

The fourth matrix is named dense *vertex-sum-connectivity matrix* and is denoted by ${}^{dv}\mathbf{SCM}$:

$$
{}^{dv}\mathbf{SCM}(T_2) = \begin{bmatrix}
0 & 2.16 & 1.86 & 3.26 & 2.03 & 2.55 & 2.66 & 2.53 \\
 & 0 & 1.16 & 1.41 & 1.15 & 1.66 & 1.66 & 2.16 \\
 & & 0 & 1.86 & 1.15 & 1.86 & 2.31 & 1.86 \\
 & & & 0 & 2.03 & 2.03 & 2.21 & 2.12 \\
 & & & & 0 & 2.41 & 2.03 & 2.03 \\
 & & & & & 0 & 2.53 & 2.55 \\
 & & & & & & 0 & 2.66 \\
 & & & & & & & 0
\end{bmatrix}
$$

It is established that the sum-connectivity indices produce structure-property-activity models comparable to the one obtained by the product-connectivity index (Randić connectivity index) (Lučić et al., 2009, 2010, 2013; Vukičević and Trinajstić, 2010; Zhou and Trinajstić, 2012). Therefore, it is expected that the use of sum-connectivity-Wiener indices would produce models comparable to those obtained by product-connectivity-Wiener indices.

### 6.2.3 Use of the Hosoya Index

The first numerical matrix, using the Hosoya index, is called the *edge-Hosoya matrix* and is denoted by ${}^e\mathbf{Z}$. This matrix was already discussed in the Section 5.12, where it was called simply the Hosoya matrix. If we sum the elements in one triangle of the ${}^e\mathbf{Z}$ matrix as originally suggested by Hosoya (1971) when he defined the Wiener index from the distance matrix, the double invariant so obtained is called the edge-Hosoya-Wiener index.

The second numerical matrix is named the *path-Hosoya matrix* and denoted by ${}^p\mathbf{Z}$. This matrix is also discussed in the Section 5.12 under the name the *dense Hosoya matrix*. If we sum up the elements in one triangle of the ${}^p\mathbf{Z}$ matrix, the index that is obtained is called the path-Hosoya-Wiener index.

The third numerical matrix is named the *sparse vertex-Hosoya matrix* and denoted by ${}^{sv}\mathbf{Z}$. An example of this matrix obtained from the sparse vertex-graphical matrix of $T_2$ is given below. Only the upper matrix-triangle is shown.

$$
{}^{sv}\mathbf{Z}(T_2) = \begin{bmatrix}
0 & 8 & 0 & 0 & 0 & 0 & 0 & 0 \\
 & 0 & 3 & 0 & 0 & 0 & 0 & 8 \\
 & & 0 & 6 & 0 & 0 & 9 & 0 \\
 & & & 0 & 7 & 0 & 0 & 0 \\
 & & & & 0 & 10 & 0 & 0 \\
 & & & & & 0 & 0 & 0 \\
 & & & & & & 0 & 0 \\
 & & & & & & & 0
\end{bmatrix}
$$

If we sum up the elements in this triangle of the $^{sv}\mathbf{Z}$ matrix, the resulting index is called the sparse vertex-Hosoya-Wiener index.

The fourth numerical matrix is named the *dense vertex-Hosoya matrix* and denoted by $^{dv}\mathbf{Z}$. An example of this matrix obtained from the dense vertex-graphical matrix of $T_2$ is given below. And again, only the upper matrix-triangle is given.

$$
^{dv}\mathbf{Z}(T_2) =
\begin{bmatrix}
0 & 8 & 6 & 10 & 7 & 12 & 13 & 11 \\
  & 0 & 3 & 4 & 3 & 5 & 5 & 8 \\
  &   & 0 & 6 & 3 & 6 & 9 & 6 \\
  &   &   & 0 & 7 & 7 & 8 & 10 \\
  &   &   &   & 0 & 10 & 7 & 7 \\
  &   &   &   &   & 0 & 11 & 12 \\
  &   &   &   &   &   & 0 & 13 \\
  &   &   &   &   &   &   & 0
\end{bmatrix}
$$

If we sum up the elements in this triangle of the $^{dv}\mathbf{Z}$ matrix, the index that results is called the dense vertex-Hosoya-Wiener index.

## 6.3   A GENERALIZED PROCEDURE FOR CONSTRUCTING GRAPHICAL MATRICES AND FOR OBTAINING THEIR NUMERICAL REPRESENTATIONS

This procedure may be generalized as follows:

1. Representation of the molecule by the corresponding hydrogen-depleted graph
2. Labeling the vertices
3. Construction of the sparse graphical matrix by consecutively removing edges or adjacent vertices from the graph
4. Construction of the dense graphical matrix by consecutive removal of paths of a given length or pairs of end vertices of paths from the graph
5. Replacing the elements of the sparse or dense graphical matrices by numerical values of the selected graph invariants to obtain the corresponding numerical matrices, that is, matrices with the numerical elements

By applying a graph invariant of choice to the numerical matrix, one can get the double invariant of a graph.

## REFERENCES

I. Gutman, Degree-based topological indices, *Croat. Chem. Acta* 86 (2013) 351–361.
I. Gutman and B. Furtula, eds., *Recent results in the theory of Randić index*, University of Kragujevac, Kragujevac, Serbia, 2008.
H. Hosoya, Topological index. A newly proposed quantity characterizing the topological nature of structural isomers of saturated hydrocarbons, *Bull. Chem. Soc. Jpn.* 44 (1971) 2332–2339.

D. Janežič, B. Lučić, A. Miličević, S. Nikolić, N. Trinajstić, and D. Vukičević, Hosoya matrices as the numerical realization of graphical matrices and derived structural descriptors, *Croat. Chem. Acta* 80 (2007) 271–276.

X. Li and I. Gutman, *Mathematical aspects of Randić-type molecular structure descriptor*, University of Kragujevac, Kragujevac, Serbia, 2006.

B. Lučić, S. Nikolić, N. Trinajstić, B. Zhou, and S. Ivaniš Turk, Sum-connectivity index, in *Novel molecular structure descriptors—Theory and application I*, ed. I. Gutman and B. Furtula, University of Kragujevac, Kragujevac, Serbia, 2010, pp. 101–136.

B. Lučić, I. Sović, J. Batista, K. Skala, D. Plavšić, D. Vikić-Topić, D. Bešlo, S. Nikolić, and N. Trinajstić, The sum-connectivity index—An additive variant of the Randić connectivity index, *Curr. Computer-Aided Drug Des.* 9 (2013) 184–194.

B. Lučić, N. Trinajstić, and B. Zhou, Comparison between the sum-connectivity and product-connectivity indices for benzenoid hydrocarbons, *Chem. Phys. Lett.* 475 (2009) 146–148.

A. Miličević and N. Trinajstić, Combinatorial enumeration in chemistry, in *Chemical modelling: Applications and theory*, Vol. 4, ed. A. Hinchliffe, Royal Society of Chemistry, Cambridge, 2006, pp. 408–472.

S. Nikolić, A. Miličević, and N. Trinajstić, Graphical matrices in chemistry, *Croat. Chem. Acta* 78 (2005a) 241–250.

S. Nikolić, A. Miličević, and N. Trinajstić, Graphical matrices in chemistry, *WSEAS Trans. Inf. Sci. Appl.* 4 (2005b) 1228–1231.

S. Nikolić, N. Trinajstić, and I. Baučić, Comparison between the vertex- and edge-connectivity indices for benzenoid hydrocarbons, *J. Chem. Inf. Comput. Sci.* 38 (1998) 42–46.

M. Randić, On characterization of molecular branching, *J. Am. Chem. Soc.* 97 (1975) 6609–6615.

M. Randić, N. Basak, and D. Plavšić, Novel graphical matrix and distance-based molecular descriptors, *Croat. Chem. Acta* 77 (2004) 251–257.

M. Randić, D. Plavšić, and M. Razinger, Double invariants, *MATCH Commun. Math. Comput. Chem.* 35 (1997) 243–259.

N. Trinajstić, S. Nikolić, D. Babić, and Z. Mihalić, The vertex- and edge-connectivity indices of Platonic and Archimedean molecules, *Bull. Chem. Technol. Macedonia* 16 (1997) 43–51.

N. Trinajstić, S. Nikolić, J. von Knop, W.R. Müller, and K. Szymanski, *Computational chemical graph theory—Characterization, enumeration and generation of chemical structures by computer methods*, Horwood/Simon & Schuster, New York, 1991, pp. 263–266.

D. Vukičević and N. Trinajstić, Bond-additive modeling. 3. Comparison between the product-connectivity index and sum-connectivity, *Croat. Chem. Acta* 83 (2010) 349–351.

B. Zhou and N. Trinajstić, Relations between the product- and sum-connectivity indices, *Croat. Chem. Acta* 85 (2012) 363–365.

# 7 Concluding Remarks

In the present book, we have discussed five classes of graph-theoretical matrices: adjacency matrices, incidence matrices, distance matrices, special matrices, and graphical matrices. A total of 170 graph theoretical matrices, which we regard as important in contemporary chemical graph theory, have been considered. They have found a wide range of applications—they are used to generate many kinds of molecular descriptors increasingly employed in molecular modeling, to generate walks and random walks, which have so far found much use outside chemistry and are used modestly in chemistry to study the complexity of molecules and chemical reactions, to generate and enumerate isomers and valence-bond structures, to start building, and to filter virtual combinatorial libraries that are so important for rational preparation of practically any desired compound. We hope that this exposition may stimulate some readers to study these matrices in more detail, since the polynomials, spectra, and properties of many of them are still poorly defined. Besides these properties, one needs to know their computational and combinatorial properties in order to establish the range of their applicability in chemistry. It is also of interest to find which, if any, of these matrices, besides the Hückel matrix (Graovac et al., 1977; Trinajstić, 1977, 1983, 1992; Hosoya, 1999, 2003) and the transfer matrix via the conjugated-circuits model (Klein and Trinajstić, 1989), may be applicable in quantum chemistry (e.g., László, 2004). A very stimulating article by Klein (2003) on graph-theoretically formulated electronic-structure theory explores these ideas. Still other graph-theoretical matrices, e.g., for the Pauling-Wheland VB (or Heisenberg) model or for the Hubard model, arise implicitly, and the wonderful world of graph-theoretical matrices is open to further, hopefully fruitful, exploration.

## REFERENCES

A. Graovac, I. Gutman, and N. Trinajstić, *Topological approach to the chemistry of conjugated molecule*, Springer, Berlin, 1977.

H. Hosoya, Mathematical foundation of the organic electron theory—How do $\pi$-electrons flow in conjugated systems? *J. Mol. Struct.* (*Theochem*) 461–462 (1999) 473–482.

H. Hosoya, From how to why. Graph-theoretical verification of quantum-mechanical aspects of $\pi$-electron behaviors in conjugated systems, *Bull. Chem. Soc. Jpn.* 76 (2003) 2233–2252.

D.J. Klein, Graph theoretically formulated electronic-structure theory, *Internet Electronic J. Mol. Des.* 2 (2003) 814–834. http://www.biochempress.com

D.J. Klein and N. Trinajstić, Foundations of conjugated-circuit models, *Pure Appl. Chem.* 61 (1989) 2107–2115.

I. László, Topological aspects beyond Hückel theory, *Internet Electronic J. Mol. Des.* 3 (2004) 182–188. http://www.biochempress.com

N. Trinajstić, Hückel theory and topology, in *Semiempirical methods of electronic structure calculations—Part A: Techniques*, ed. G.J. Segal, Plenum Press, New York, 1977, pp. 1–27.

N. Trinajstić, *Chemical graph theory*, Vols. I and II, CRC, Boca Raton, FL, 1983.

N. Trinajstić, *Chemical graph theory*, 2nd ed., CRC, Boca Raton, FL, 1992.

# Index